An Introduction to Usability

PATRICK W. JORDAN

PHILIPS DESIGN

UK   Taylor & Francis Ltd, 11 New Fetter Lane, London, EC4P 4EE
USA Taylor & Francis Inc., 325 Chestnut Street, 8th Floor, Philadelphia, PA 19106

Reprinted 2001

Taylor & Francis Ltd, is an imprint of the Taylor & Francis Group

Copyright © Taylor & Francis Ltd 1998

British Library Cataloguing-in-Publication Data
A catalogue record for this book is available from the British Library
ISBN 0–7484–0762–6 (paperback)
ISBN 0–7484–0794–4 (cased)

Library of Congress Cataloging in Publication Data are available

Cover design by Hans Jacobs and Robin Uleman (Philips Design)
Photographs by Jim Cockerille
Typeset in Times 10/12pt by Graphicraft Typesetters Ltd, Hong Kong
Printed by T.J. International Ltd, Padstow, UK

Contents

Acknowledgements

In addition to the publications cited in this book, much of this text draws on my own personal experiences of working as a researcher and practitioner on usability-related projects. I am grateful to colleagues from the University of Glasgow and from Philips Design – I have learned a great deal from them over the last six years or so. In particular Graham Johnson, Steve Draper, Eddie Edgerton, Paddy O'Donnell, Ian McClelland and Fred Brigham deserve a mention.

Similarly, I owe a debt of gratitude to many fellow professionals from other universities, companies and research institutions with whom I've been privileged to have contact, through conferences or just informally. Again, this has been a valuable learning experience.

Thanks to those who have encouraged me during the process of writing this book, including all at Psychology Press especially Luci Allmark.

CHAPTER ONE

Introduction

USABILITY

The products that we use in our homes and at work are becoming ever more complex in terms of the features and functionality that they contain. In order that users can benefit from these features it is important that those responsible for product creation ensure that the requirements and limitations of those using the products are taken into account. Indeed, users are beginning to demand this. Whilst at one time usability problems may have been the price that users were prepared to pay for 'technical wizardry', this no longer seems to be the case. People are increasingly unwilling to tolerate difficult to use products. Manufacturers have recognised this and increasingly products are being advertised as 'user friendly' or 'ergonomically designed'. Thus, usability is becoming seen as an important issue commercially.

Usability issues have received increasing attention over the last few years. This is manifest in a number of ways. One is the expanding literature relating to usability, including journals, books, and even magazine and newspaper articles. There are also a number of international conferences and seminars dedicated to usability-related subjects. Examples include the Ergonomics Society Conference in the UK and the Human Factors and Ergonomics Society Conference in the USA. However, perhaps the most important reflection on how seriously usability issues are now being taken is the sharp increase in the number of professionals employed by industry who are charged with ensuring that products are easy to use. These include human factors specialists and interaction designers. In addition, product designers and software programmers are increasingly expected to have an awareness of usability issues and to put the user at the centre of the design process.

Indeed, usability may be one of the few areas left to manufacturers where it is possible to gain a strong commercial advantage over the competition. Manufacturing processes have now reached a stage of sophistication whereby any possible advantages in terms of manufacturing quality or cost savings are likely to be marginal. Offering customers 'user-friendly' products could be seen as something new in markets where the technical and functional specifications vary little between brands – effective human factors input can, then, mark out a product as significantly preferable to others on the market. Of the factors involved in the product creation process, then, usability issues can be amongst the most significant in terms of influencing the commercial success of the product.

Aside from the commercial implications, lack of usability can have effects that range from annoying the users to putting their lives at risk. Whilst lack of usability in a video cassette recorder (VCR) may result in the user recording the wrong television programme, lack of usability in a car stereo may put lives at risk by distracting drivers' attention from the road.

Usability is not only important with respect to consumer products, but also those used in a professional or commercial context. Nearly all professional jobs now involve the use of computers. This has meant that over the last few years entire workforces have had to become 'computer-literate' or rather that computers have had to become 'user-friendly'. Unless these computer applications are made usable the potential benefits of computerisation, in terms of the effectiveness and efficiency with which businesses can be run, will not be fully realised. Indeed it has been estimated that in the 1980s, in computerised offices, up to 10% of working hours were wasted solely due to usability problems (Allwood, 1984).

Similarly, usability can have important implications for productivity in manufacturing environments. If operators have difficulties with machinery they will not be able to produce as much in a given time and, in all probability, it will be of a lower quality than it might otherwise be.

As mentioned previously, usability issues can have a major effect on safety. Consider, for example, a product for use in emergencies such as a fire extinguisher. This must be easy to use at the first attempt in situations where the user may well be under considerable stress. Clearly, there may be serious consequences if the user has to struggle with it. Similarly, lack of usability has been shown to have been a causal factor in many industrial, domestic and transport accidents.

The importance of usability will be expanded at some length in Chapter 2.

AIM AND CONTENTS OF THE BOOK

The aim of this book is to provide the reader with an introduction to the issue of 'usability'. In the next chapter, the reader will be introduced to the concept of usability – what usability means in 'lay person's terms' and how it has been defined formally and operationalised.

The third chapter will look at the principles of designing for usability – the characteristics that separate usable designs from those that are not usable. Chapter 4 outlines the elements of a design process which are required in order to guarantee that usability issues are adequately tackled.

The fifth and sixth chapters concentrate on usability evaluation. In Chapter 5, a series of methods for evaluating usability are outlined. Chapter 6 then describes what is required in order to conduct an effective usability evaluation.

SCOPE OF THE BOOK AND LIMITATIONS

It is hoped that readers of this book will emerge with an overview of what usability is and have a basic understanding of what must be taken into account in order to design for usability and conduct usability evaluations. No previous knowledge of usability issues is assumed and the book should be accessible to all, regardless of academic or professional background.

This is not a 'manual'. Reading the book will not give a person new skills – for example, it will not turn a marketing professional into someone who can design usable products or a computer scientist into an evaluation specialist. However, it should give all readers an awareness of the issues and give those who already have particular skills the knowledge to enhance their effectiveness. For example, those who currently conduct usability evaluations will find many new techniques to add to their repertoire, whilst those with experience of product or software design should be able to apply to their work the ergonomic design principles outlined.

WHO IS THE BOOK FOR?

The book was written primarily for students and those whose profession is connected with the process of product creation. 'Product', as used in this book, is a generic term covering, for example, computer software, consumer products, products for use in the professional environment and manufacturing equipment/machinery. Courses for which the book should be suitable include: human–computer interaction, product design, human factors/ergonomics, information technology, marketing, multimedia and modules on human–product interaction in psychology, computer science and engineering courses. Professions at which the book is aimed include human factors, interaction design, product design, software design, marketing, product management and engineering.

CHAPTER TWO

What is Usability?

In the previous chapter, the concept of usability was introduced. In this chapter usability will be defined, its various components will be introduced and the various measures by which usability can be quantified will be discussed. Examples will be given throughout, demonstrating the importance of usability with respect to a variety of different products, tasks and situations.

CONCEPT AND DEFINITION

Informally, usability issues can be thought of as pertaining to how easy a product is to use, i.e. they are to do with the 'user-friendliness' of products. More formally, the International Standards Organisation (ISO) defines usability as '. . . the effectiveness, efficiency and satisfaction with which specified users can achieve specified goals in particular environments' (ISO DIS 9241-11).

Effectiveness refers to the extent to which a goal, or task, is achieved. In some cases, the distinction between a task being achieved successfully or not may simply be success or failure in that task. For example, if when using a word processing package, the user decides that he or she wishes to open a new file, then the result of attempting such a task can have only two meaningful outcomes – success, whereby the new file is created, or failure, whereby it isn't. However, there are situations where effectiveness can be measured in terms of the extent to which a goal is achieved. Consider, for example, a production process involving, say, a computerised-numerically-controlled (CNC) milling machine. If the machine operator's goal were to produce 100 components per day, then if they were able to produce 80 per day, this might be seen as an effectiveness level of 80%.

Efficiency, meanwhile, refers to the amount of effort required to accomplish a goal. The less effort required, the higher the efficiency. Effort might be measured, for example, in terms of the time taken to complete a task or in terms the numbers of errors that the user makes before a task is completed. Consider again the example of creating a new file with a word processing package. If the user is able to create the new file instantly at the first attempt, then the package would be considered more usable than if the user had to spend a couple of minutes thinking what to do, or than if he or she had activated several inappropriate commands before eventually finding the right one. The difference in usability,

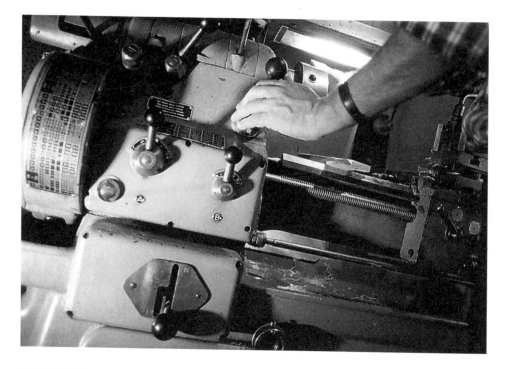

Figure 2.1 With work–related products, effectiveness and efficiency may be the most important aspects of usability, whereas with leisure products, the satisfaction aspect may be the priority.

in this case, would be a difference in efficiency but not in effectiveness. In both cases users have achieved the goal, but the more time taken or the greater the number of errors made the less the efficiency. Similarly, the issue of efficiency can also be considered in the context of milling components. If the operator has had to work non-stop, concentrating very hard in order to mill 80 components, then the machine could be said to be less efficient than if the operator had had time to drink coffee and chat with his or her workmates whilst making the 80 components.

Satisfaction refers to the level of comfort that the users feel when using a product and how acceptable the product is to users as a means of achieving their goals. This is a more subjective aspect of usability than effectiveness or efficiency. It may also be more difficult to measure. This, however, does not mean that it is inherently any less important than the other two. Indeed, there may be many situations where it is the most important usability consideration. Consider, for example, the case of television usage. Many high-end televisions have a vast array of functions, often controlled via on-screen menu systems. Designing for usability is, then, very important in this context. However, the users' level of satisfaction when operating the television may not always be directly linked to the levels of effectiveness and efficiency with which they complete tasks – issues such as the 'look and feel' of the interface may also play an important role. For example, if users can achieve their goals, such as adjusting the brightness of the picture or altering the levels of bass or treble in the volume, more quickly with TV 'A' than with TV 'B', but prefer the visual layout of the menus on TV B, then they may find TV B more satisfying overall to use. As a TV is something designed to bring pleasure to users, rather than as a tool with which to achieve a particular level of productivity, then satisfaction is likely to be the most important aspect of usability. Of course, satisfaction *will* often be strongly correlated with effectiveness and efficiency. After all, if it took too much effort to adjust TV B or if the users were often unable to use the controls then it seems unlikely that they would be satisfied with it. Nevertheless, satisfaction is a conceptually separate aspect of usability and may be, at least partially, independent of the other two aspects.

In general, satisfaction might be seen as the most important aspect of usability for products whose use is voluntary. For example, if users of consumer products do not find them satisfying to use, they do not have to use them. Perhaps more importantly from a commercial point of view, they do not have to buy any more products from the same manufacturer. Indeed, usability is something that consumers are now expecting from a product. Whilst at one time it was accepted that high tech products were complicated and usability was seen as a bonus, lack of usability is now seen as a major source of discontent with consumer products. Conversely, in situations where people are 'forced' to use products, such as manufacturing equipment or products used in professional environments (for example software packages), it might be argued that effectiveness and efficiency are at least as important – at least from the point of view of the employers!

An important point to note about the ISO definition of usability is that it makes clear that usability is not simply a property of a product in isolation, but rather that it will also be dependent on who is using the product, the goal that they are trying to achieve and the environment in which the product is being used. Usability is, therefore, a property of the interaction between a product, a user and the task, or set of tasks, that he or she is trying to complete. In Chapter 3 some of the properties of a product that affect usability will be outlined. Examples of the effects on usability of task and context of use will be given throughout the book. Meanwhile, the effects of user characteristics are discussed in detail below.

The effect of user characteristics on usability

A product that is usable for one person will not necessarily be usable for another. There are a number of user characteristics which can be predictors of how easy or difficult a product is to use for that person. Designing for usability means designing for those who will use the product in question. It is vital, then, to have an understanding of who the users of the product will be and their characteristics.

Experience

Previous experience with the product itself is likely to affect how easy or difficult it is to complete a particular task. Clearly, if a user has performed the task with the product before, then he or she is likely to find the task easier on subsequent attempts. This issue is considered in more depth later in the chapter in the section 'The Components of Usability' which describes how a product's usability profile changes as the user becomes familiar with using it for a particular set of tasks. Similarly, experience with using a product for one task can also affect how easy or difficult users find it to complete other tasks with the same product. For example, if the user of a statistics program has previously used the package to perform, say, a correlation statistic, then when he or she tries to use the package to perform a significance calculation, for instance, it may be possible to draw on experience gained from performing the correlation. It might be that this previous experience enables users to guess which menu the appropriate command will be on, or in the case of a command line driven interface, the order in which they must type in the arguments. With well-designed products, users should be able to generalise from previous experience to help them complete new tasks. However, if the product is not well designed, then users may have difficulties in trying to generalise. Consider again the statistics package. If the correlation commands were placed, say, at a different level in the menu structure to the commands for significance calculations then the users' previous experience might be of little help. Indeed it might actually be harder for users to complete the significance calculation than if they had not previously performed the correlation calculation, as there is the danger that by drawing on this experience they will go to an inappropriate level in the menu structure. Similarly, with a command line driven package, if the argument order differs for correlations and significance calculations, then generalising from one to the other will also cause problems. This principle is known as consistency. Consistency in a product's interface enables the user to successfully generalise from one situation to another. Consistency, along with the other major principles of usable design, is discussed in depth in Chapter 3.

Experience with other similar products will also affect how usable a new product is for a user. Again, the issue is one of being able to generalise, but this time between products rather than between tasks. For example, if someone has experience of using a particular word processor for formatting documents, then he or she will be able to draw on this experience when formatting a document with a word processor that is new to him or her. As in the case of experience of other tasks with the same product, experience of other products may not always be helpful. It is only likely to be helpful if the design of the new product enables the user to draw on these experiences. For example, if a new word processor were to employ a totally different method of document formatting to any other package on the market, then experience with other word processing packages is unlikely to be helpful to the user. This issue can sometimes lead to something of a dilemma for the designer of a new product. Obviously, there are currently many unusable products on the market and radical improvements in design are necessary to bring about improvements. However, with radical changes the inherent usability benefits of compatibility with other products may be lost.

Domain knowledge

Domain knowledge refers to knowledge relating to a task which is independent of the product being used to complete that task. For example, when using a spreadsheet package on a computer, users' performance may be affected by how much knowledge they have of spreadsheets generally. Someone who knows how a spreadsheet is laid out may be expected to perform better with the package than someone who has no idea about spreadsheets. Again, the designer can take this into account and try to create a package that will make use of any domain knowledge, whilst taking care not to *assume* that users will have this knowledge, as this might make the package difficult for those without knowledge of the domain.

Cultural background

The cultural background of the users can also influence how they interact with products. This is because of 'population stereotypes' that people hold. In the USA and mainland Europe, for example, switches are typically flicked upwards in order to turn something on. In Britain, conversely, it is typical to flick a switch down to turn something on. Similarly, emergency exit signs in the USA have red lettering, whilst in Europe the lettering is green.

Population stereotypes should be taken into account when designing for particular markets. This is particularly so when products have a safety critical aspect to them, because in the heat of an emergency people may revert to 'instinctive' behaviours. So, for example, if designing potentially hazardous industrial machinery for sale to Britain, the switching on of the machine should require the operators to switch it upwards in the event of an emergency shut-down being necessary. However, if the machine was for sale to companies in continental Europe or the USA, then the switching should be down for off.

Another issue to consider when designing for different cultures is that there are often differences in people's physical characteristics that are related to race and nationality. For example, the average Japanese person is smaller than the average European. This can have implications in, for example, the design of car interiors. Manufacturers selling to both markets must take the physical dimensions of both populations into account when considering issues such as the range of adjustability required in seating or the position of the controls. Clearly, if controls are placed too far away from the driver's seat, they may be unusable for smaller people.

Machinery guarding is a safety critical area where accounting for the physical dimensions of people is very important. Consider a finger guard on a pressing machine, for example. These guards have to be designed to allow the workpiece to fit underneath, yet prevent the operator from getting his or her fingers through. Again, it is vital to consider who will be operating these machines. If they are designed to prevent the fingers of male Europeans from fitting underneath, this does not necessarily mean that they will be safe for, say, female Asians.

Disability

Clearly, products that are usable for those who are able-bodied will not necessarily be usable for disabled people. However, by paying attention to the needs of those with disabilities, it is possible to provide opportunities for the disabled which might otherwise be restricted. An example is the issue of wheelchair access to buildings and the potential for free movement within them. A building with ramp access, automatic doors and lift access to all floors will be far more usable for a wheelchair-bound person than one with steps to the entrance, spring-loaded swing doors and only stairway access to floors.

New technology should also enable some products to be made more usable for those with special needs. For example, speech recognition interfaces offer possibilities for those without the use of their hands. These are becoming available in the context of an increasing number of applications. For example, it is now possible to design telephones which can be operated without the use of a handset and without the need to use keys to dial the number. Such phones allow the user simply to speak the number to be called and the telephone will dial this automatically. The user can then converse without the need to hold a handset to his or her mouth and ear, as the phone can pick up the user's voice via a long-range microphone and amplify the other person's voice via a speaker.

Products that are designed to be usable for the disabled can also benefit able-bodied users. For example, hands-free telephones can have advantages over conventional mobile phones when fitted in cars, as they allow drivers to carry on conversations without removing a hand from the steering wheel.

Two products that were originally designed for the disabled, but which have found wide acceptance amongst the able-bodied are ball-point pens and TV remote controls. With traditional ink pens, users without full control of their hands would often break the delicate nibs. The ball-point is a far more robust design, requiring a lesser degree of control to operate. The remote control, meanwhile, was originally designed for those who had difficulties in moving from their chairs to the television set in order to change the channel or alter the settings. These, then, are examples of how designing for the disabled can provide benefits for all.

Age and gender

There are user characteristics that will often vary with age and gender which need to be taken into account when designing certain products for usability. For example, young men are, on average, physically stronger than women and the elderly. One area in which this has implications is in the design of cars. The wider use of power-assisted steering, for example, means that many larger vehicles can now be driven by a greater percentage of women and elderly than would previously have been the case. Similarly, vehicle handbrakes were often too stiff for many women and elderly people to put on full lock. Again, power assistance has now made these operable for a wider range of drivers.

There are also attitudes which people may be more likely to hold according to their age or gender. There are, for example, respects in which men and women may be treated differently in childhood which can lead to them forming different attitudes towards, say, high tech products. It may be that boys are brought up to pay more attention to scientific subjects, such as mathematics, physics and computer science, whereas girls are encouraged to concentrate on the arts subjects, such as literature and languages. This may lead to males feeling more confident when faced with complex-looking products and thus they may place a high emphasis on issues such as functionality. Females, on the other hand, may be more concerned with issues such as the aesthetics of the product. This may mean that the satisfaction component of usability will be affected by the appearance of the product. Young men may be more inclined than most other sections of the population to prefer products than have a 'high tech' image over, say, products with a 'softer' form language.

Similarly, age can be a factor, because different generations have grown up with different types of technology. Whilst the younger generations are likely to have grown up with a high exposure to computers, this may not be the case for older people. This means that older people may be less accepting of computer-based products and may be deterred from using them. Younger people, however, may not be put off by this type of product. Thus, when designing for older people, it may be preferable to 'disguise' the computerised element of products. For example, older people may prefer video cassette recorders (VCRs) with which they simply have to type in a number in order to program the machine to record. Younger users, though, may be happier to go through a comparatively complex programming sequence.

THE COMPONENTS OF USABILITY

Apart from the issue of 'experience', the user characteristics discussed above may be thought of as comparatively stable. They are characteristics that, if they change at all, will probably change over a comparatively long period. However, users' performance with a product is likely to improve significantly in relation to tasks which they repeat with the product over time. Thus, the usability of a product for a particular person completing a particular task may change very quickly as the task is repeated. What might once have been terribly difficult may five minutes and six repetitions later be very easy. It may just take a little practice to 'get the hang of' a product. To reflect this, Jordan et al. (1991) developed a three-component model of usability accounting for the change in level of task performance with repetition. The components are guessability, learnability and experienced user performance (EUP). These are associated with, respectively, first time use of a product for a particular task, the number of task repetitions required until an acceptable level of 'competence' is reached, and the relatively stable level of performance that an experienced product user reaches.

A later extension of the model (Jordan, 1994a) includes an additional two components – system potential and re-usability. These are concerned with the theoretical optimal performance obtainable with a product with respect to particular tasks and the level of performance achieved when a user returns to a task with a product after an extended period of non-use. These five usability components are explained below. Definitions of each component are given, based on the ISO definition of usability.

Guessability

Guessability is a measure of the cost to the user in using a product to perform a new task for the first time – the lower the cost (for example, in terms of time on task or errors made) the higher the guessability. Guessability is likely to be particularly important for products that have a high proportion of one-off users, for example fire extinguishers, door handles on public buildings, or public information systems. Lack of guessability can have serious commercial implications as it may put users off products which might have been comparatively easy to use in the longer term. For example, a customer may choose to buy the stereo which was easiest to use at the first attempt in a shop, even though this may not be the stereo which would have been easiest to use in the long term.

Guessability is of less importance in situations where product operation is initially demonstrated to the user, where the user has training with a product, or where there is no pressure to complete tasks successfully at the first attempt. Examples might include aircraft control panels or military equipment, which are typically designed to be used by experts after considerable training. Note, however, that even with these complex products there may still be particular tasks which must be guessable. For example, a pilot may have to make an emergency problem diagnosis or error recovery. Even though he or she may be experienced with the aircraft generally, there may still be rare tasks that the pilot has not encountered before. A parallel example concerns users who are generally experienced with a particular software package. Again, there may be rare tasks which they have not performed before for which they need to find the appropriate command. This command should, then, be guessable. Menus can offer advantages over command line interfaces in this respect as the user may have seen the required command during earlier interactions with the interface and will thus know where it is when needed.

Guessability

The effectiveness, efficiency and satisfaction with which specified users can complete specified tasks with a particular product for the first time.

Learnability

Learnability is concerned with the cost to the user in reaching some competent level of performance with a task, but excluding the special difficulties associated with completing the task for the first time. If the method for performing a task proved easily memorable after the first completion, the product would, then, be highly learnable for this task. Learnability can be particularly important in situations where training time is short, or where a user is to be self-taught with a product. Consider, for example, the case of temporary secretaries using word processors. The nature of temporary employment means that these employees may be moving from job to job fairly regularly and that they will have to get used to using a variety of different word processing packages. It is also likely that employers will expect secretaries to adapt to new packages quickly and with little or no formal training. Clearly, if they do not reach a competent standard with the word processors fairly quickly a significant proportion of their working time will be wasted. Learnability, then, would be an important issue here. In the context of computer software, command line interfaces which rely on users learning and recalling command names might be expected to be less learnable than recognition-based interfaces – for example graphical or menu-based interfaces. Recognition-based interfaces tend to be more learnable as users can operate them without having to retain detailed information about the interface in their heads. Rather, visual cues prompt the user at the time of use (Mayes et al., 1988).

Learnability will be less important in situations where there are comparatively few restrictions on training time and resources. Again, a pilot learning to fly an aircraft would be an example of this.

The term 'learnability' has been widely used throughout the usability literature. However, it has often been given a different meaning than that described here. Indeed, different practitioners appear to give the term different meanings. Payne and Green (1986) use the term with reference to completing new tasks for the first time (here referred to as guessability) whilst others, such as Shackel (1986, 1991) use the term more generally, to refer to a product's usability for any user who could not be deemed experienced. Certainly, the idea that there is a distinction between usability for an experienced user and a user with less experience is not unique to the five-component usability model. However, the five-component model perhaps outlines a more concrete approach to the issues involved – for example by paying attention to the guessability/learnability distinction and by supporting judgements as to when a user can be deemed to be 'experienced' as opposed to 'inexperienced'. For example, many studies reported in the literature, described as studying new users, either disregard the time to complete tasks for the first time, or give subjects special help at this stage. This means that they omit measures of guessability.

Learnability

The effectiveness, efficiency and satisfaction with which specified users can achieve a competent level of performance on specified tasks with a product, having already completed those tasks once previously.

Experienced user performance (EUP)

This component of usability refers to the relatively unchanging performance of someone who has used a product many times before to perform particular tasks. Although performance may not always level off at an asymptotic level, there will probably come a stage with most products where significant changes will only occur over comparatively long time-scales. This is in contrast to the learnability phase where changes occur comparatively rapidly.

EUP will be the most important component of usability where there is comparatively little pressure to learn quickly, but where it is important that once product operation has been learned, users then perform at a high level. The use of specialist software packages, for example for computer-aided design (CAD), might come into this category, as would flying an aircraft or driving a car.

EUP will be of less importance for products which are likely to be used either intermittently or on a one-off basis. An example would be computerised tourist information systems that can often be found in British towns and cities. With these, visitors to the town may want to look up, say, a selection of hotels that are available in the vicinity or information about the local restaurants. They may look up three or four items of information, but are unlikely to use the system over a protracted period. If such a system could only be used quickly and easily after interacting, say, twenty or thirty times with it, this may be of little benefit. What would be important is whether or not users of the system could find the information they wanted on the few occasions that they used it – thus guessability and possibly learnability would be the components of interest here.

EUP

The effectiveness, efficiency and satisfaction with which specified experienced users can achieve specified tasks with a particular product.

System potential

System potential represents the maximum level of performance that would be theoretically possible with a product. It is, then, an upper limit on EUP. Consider, for example, a command line interface to a software program. The number of keystrokes required to complete a particular task would be a measure of the system potential of the product for that task – the less keystrokes the higher the system potential. It would not matter, then, how experienced or sophisticated a user became with respect to the product, he or she would still have to complete a certain number of keystrokes in order to achieve a task.

Often, the EUP of a product will fall short of its system potential. This may be because the user has never learnt the optimal method of task performance or, perhaps, because he or she has not had the opportunity to perform tasks on the upper level of performance. An example, in the domain of computers, of a user sticking with a non-optimal method of interaction might be a menu-driven interface with the option to invoke some commands using accelerator keys as an alternative to picking a command with the mouse. Sometimes experienced users of these types of interface will not bother to learn how to invoke commands via the accelerators, but will stick with the slower alternative of selecting from the menus.

An example of a case in which users are rarely given the opportunity to exploit the full system potential is that of driving a car. The top speeds of most cars are far in excess of the speed limits for driving on public roads. Thus, although it might be theoretically possible to complete the task of getting from one end of a city to the other at an average speed of 100 mph, the law (not to mention considerations of road safety) does not permit this. Interestingly though, market research suggests that top speed, one measure of a car's system potential, is a major consideration for many in deciding which vehicle to buy.

Over extended periods of time it may be possible that what appears to be a steady level of experienced user performance will move closer to system potential. This might happen because a user suddenly discovers a new technique or approach that facilitates a step jump in their performance. Return to the example above – the menu-based interface with accelerator options. It may be that after a protracted period of only using the menu commands, a user suddenly decides that they will use the accelerators to invoke certain commands. The EUP associated with these commands, then, is likely to show a step increase, perhaps to the level of system potential. This phenomenon is referred to by Norman, Draper and Bannon (1986) as 'shells of competency' – by taking a new, more efficient, approach the user has broken through the 'shell' which represented the limit on his or her performance on a task.

System potential will be important when it is the limiting factor on EUP and less important where it isn't. With command line interfaces, for example, long keystrokes may make interaction cumbersome even if users no longer have trouble remembering what they have to type – the system potential, then, is likely to restrict EUP. Conversely, in the case of, for example, an author using a word processing package to write a book,

EUP is unlikely to be affected by the package's system potential. The time taken to complete the book is far more likely to be dependent on the speed at which the author can think and type, rather than the time the word processing package takes to respond to keystrokes.

System potential

The optimum level of effectiveness, efficiency and satisfaction with which it would be possible to complete specified tasks with a product.

Re-usability

This component of usability refers to the possible decrement in performance after the user has not used the product for a comparatively long period of time. This decrement could occur, for example, because the user has forgotten how to perform a particular task with a product or because he or she has forgotten the function of a control or, perhaps, because the user has forgotten where the required control is located.

When considering re-usability the issue of what constitutes a 'comparatively long period of time' away from performing a particular task with the product is one that needs to be addressed. The learnability and EUP components of usability assume that the user is using a product for a particular task fairly regularly. If there is a gap between repetitions of a task that is significantly longer than that which usually would be expected between task repetitions then the user's performance might now be said to be a reflection of the system's re-usability. Jordan (1994a) cites the example of a researcher using a statistics package. Typically, after a period of data gathering, a researcher may spend several days using the statistics package to analyse the data – this may involve repeating particular tasks with the package several times over the course of those days. However, after the analysis is complete, the chances are that the researcher won't use the package again until another study is completed – possibly a period of several months. When returning after such a period away, the level of performance achieved on the tasks could be considered to be a measure of re-usability. This example – where a product is used in intermittent 'bursts' – is typical of situations in which re-usability is likely to be important.

It should be noted that, as with each of the other components of usability, re-usability refers to performing a specific task. Thus the 'comparatively long period of time' does not necessarily refer to a time away from a product as a whole but merely a particular task for which the product can be used. For example, the user of a VCR may, over a period of time, continue to use the basic functions (such as 'play', 'fast forward' and 'rewind') on a daily basis, yet may not, during this period, use the programming functions. When he or she returned to using these functions the re-usability of the VCR with respect to programming would be important.

Clearly, the level of performance achieved may be dependent on the length of time away from the interface and in order that measures of re-usability are meaningful, decrement in performance must be considered in the light of this length of time. Nevertheless, there may be some tasks for which even a short time away brings the user back to the start of the learning curve and other tasks for which performance would remain at or near EUP even after a long time away. For example, riding a bicycle is a skill that, according to many, is never lost once learned.

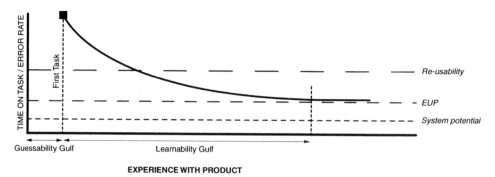

Figure 2.2 Idealised learning curve illustrating the five components of usability.

Re-usability

The effectiveness, efficiency and satisfaction with which specified users can achieve specified tasks with a particular product after a comparatively long period away from these tasks.

Each of the components of usability is loosely associated with a different section of a notional learning curve. An example of such a curve is illustrated in Figure 2.2.

WHY USABILITY IS IMPORTANT

In the introductory chapter some examples were given of why usability is important. More examples are given here. Lack of usability can cause problems which, at one end of the scale, may frustrate or annoy the user and, at the other end of the scale, might be life-threatening.

Annoyance

All of us must have experienced the frustration of coming across products in our daily lives that we can't use or which cause us difficulties. In his classic book *The Psychology of Everyday Things*, Donald Norman (1988) reports that many people he has spoken to have difficulties with everyday items such as washing machines, sewing machines, cameras, VCRs and cookers (turning on the wrong burner). Products such as these are intended for people's convenience and enjoyment – if they are difficult to use they cause annoyance and frustration and defeat their intended purpose.

Financial implications (product sales)

Whilst users may once have accepted lack of usability as the price to pay for extensive functionality in products, this situation is changing. As public awareness of usability issues increases, usability is becoming a factor in purchase decisions. This is reflected both in the use of slogans such as 'Ergonomically Designed' in advertisements and the increasing number of human factors specialists that are employed by companies manufacturing consumer products (Jordan, McClelland and Thomas, 1996).

Design issues – including usability – may be one of the few areas left where manufacturers can gain significant advantages over their competitors. In many cases manufacturing processes have now become so sophisticated and standardised that any further advances made by one company only provide marginal benefits in terms of, say, cost and reliability of the product. However, because comparatively little attention has been paid to usability in the past, those who do consider these issues when designing products can develop products that are far easier to use than other similar products on the market. For example, the success of direct manipulation computer operating environments, such as the 'Apple Macintosh' and 'Windows' is probably largely due to the usability benefits these environments bring as compared to 'command line' operating environments.

Financial implications (productivity)

Unusable products in the working environment waste time and with it money. A field study by Allwood (1984), for example, indicated that difficulties in using computers could cost companies between 5 and 10% of total working time. The same principle holds true for industrial machinery. Consider, for example, a lathe or a milling machine that is used for making components. The more difficult these are to operate the fewer components an operator is likely to be able to manufacture in the course of the day.

The usability of products used in the workplace can also have an effect on the level of job satisfaction amongst that organisation's employees, particularly in cases where a significant proportion of an employee's time is spent using a particular product. If employees do not have a high level of job satisfaction the organisation can expect a higher rate of absenteeism, a higher turnover of staff and lower motivation amongst the workforce – all factors that are likely to have a negative effect on productivity.

Safety

In some cases the usability of a product can affect the safety of those using the product, as well as the safety of others. Consider, for example, an in-car stereo system. Car stereos are typically operated by the user whilst driving the vehicle. If the stereo is difficult to use, this is likely to distract the user from driving the car – for example because the user must take his or her hand off the steering wheel or remove his or her gaze from the road. Of course the consequences of distracting attention from the driving task are potentially disastrous. This matter is one which is receiving increasing attention in the light of the development of in-car systems for providing information to drivers.

There are also many domestic situations where lack of usability could be dangerous. Consider the classic ergonomics problem of matching the knobs on the front of a cooker to the burners that they control. If, for example, someone was cooking using two or three burners and then turned the wrong one off when removing a pan, a hot surface would be left exposed.

Perhaps, however, the areas which cause greatest safety concerns are in the design of control rooms for the control of potentially hazardous operations. Norman (1988), for example, outlines a number of design failings that contributed to the nuclear power plant disaster at Three Mile Island in the USA. Here, the disaster was blamed on 'human error', as the operators failed to diagnose what the problems with the reactor were. However, Norman argues that there were serious design faults that led to the misdiagnosis taking

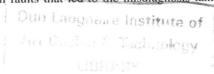

Figure 2.3 Driving is a safety critical task. It is important that products designed for use in the in–car environment do not distract a driver's attention from the road.

place. For example, one of the crucial instruments that the operator needed to check was actually on the back of the control panel! Similarly, human factors issues may have played a crucial role in the Piper Alpha oil rig disaster. In a paper presented to the annual conference of the British Ergonomics Society Gibson and Megaw (1993) claim that '. . . poor design has a large potential significance in terms of the disaster'. One of the major concerns was that items of equipment which, because of their functions needed to be operated in sequence, were not located within convenient distances of each other.

MEASURES OF USABILITY

The ISO definition of usability mentions three separate aspects – effectiveness, efficiency and satisfaction. Each of these can be translated into fairly concrete measures. Some measures associated with effectiveness, efficiency and satisfaction are outlined below. Definitions of effectiveness, efficiency and satisfaction are given in the subheadings as a reminder (based on ISO DIS 9241-11).

Effectiveness: the extent to which a goal, or task, is achieved

Task completion

The most basic measure of whether or not a product is effective for a particular task is whether or not the user can complete that task with the product. For example, if the product were a vacuum cleaner and the task were to clean dust from a carpet, then the cleaner could be regarded as effective for that task if the user could get the carpet clean.

Similarly, a microwave oven could be regarded as effective for cooking pies if users were able to successfully cook pies using the microwave – if cooking pies was too difficult for the users to manage then the microwave would be ineffective for this task.

Often, when tasks are looked at individually, effectiveness can be a 'black and white' issue – either a product-user combination is effective for a task or it isn't. Consider a simple product such as a kettle. This has two basic functions – boiling water and facilitating the transfer of the water, when boiled, to some other receptacle, such as a cup or a pan. If the user was able to use a kettle to boil water and successfully transfer this water to the appropriate receptacle, the kettle would be totally effective for its intended task; if not, then the kettle would be totally ineffective.

With more complex systems it is possible that a user completes a task with only partial success. Return again to the example of preparing a letter on a word processing package. This task could be broken down into a number of sub-tasks. Jordan (1992a) reports an empirical comparison of the direct manipulation and command driven versions of the same word processing package. For purposes of analysis the tasks set to participants in each of the evaluations were broken down into 12 separate sub-tasks. Examples of the sub-tasks were: changing the format of a section of text, changing the font size of a section of text, deleting some text from the letter and inserting some text into the letter. In this context it might be reasonable to consider the degree of effectiveness of the word processing package for preparing a letter in terms of the number of sub-tasks that the user was able to successfully complete. After all, if they were not able to, for example, change the font size of the text, this would not mean that the other sub-tasks that they had been able to complete would be valueless.

Quality of output

With some products it is possible that a user is able to complete a task with a product, but that the output resulting from the completion of a task is of variable quality. Manufacturing machinery is an example of this. Consider an operator manufacturing components using a lathe. Clearly, the dimensions of these components must be within certain tolerances in order for them to be of any value. However, within these limits there will probably be ideal dimensions to which the components should conform. Therefore, the closeness of the component's dimensions to these ideal dimensions could be used as a measure of effectiveness.

Efficiency: the amount of effort required to accomplish a goal

Deviations from the critical path

For most tasks there is a critical path for task performance – the method of approaching the task that requires the least effort (i.e. the least steps or the least time to complete). If the user deviates from the critical path then this is negative in terms of efficiency. For example, consider the user of a menu driven computer-based package, such as a spreadsheet program. Imagine that the user wanted to save his or her work and that this required the invocation of a command called 'Save' that was placed on a menu headed 'File'. If the user were first to open other menus – with headings such as 'Edit' or 'Utilities' – before opening the 'File' menu and selecting 'Save' this would represent a deviation from the

Figure 2.4   Users often find it irritating and time–consuming to have to consult a manual in order to complete a task. It could be argued that if a product is well designed from the usability standpoint its operation principles should be self-evident and a manual would thus be unnecessary.

critical path. There would be a cost in terms of time and effort associated with this deviation. It could be argued that deviations such as these are not actually errors because they don't require any corrective action. In this example, the user has not actually invoked any other command, so nothing in their word processed document has been changed.

Having to consult a manual or (in the case of computer-based products) an on-line help system could also be classified as a deviation from the critical path. Although most products come with supporting literature such as manuals or help systems, the role of these should be minimal if a design is to be usable – certainly providing a manual should not be seen as an alternative to good design. Generally, it would be expected that the more usable a product were, the less frequently the user would have to consult the manual. Certainly, users don't like having to refer to documentation in order to use a product. In a newspaper interview, the chairperson of Apple Computers claimed that one of the reasons for the phenomenal success of direct-manipulation interfaces was that users could achieve their goals '. . . without having to read a 500 page instruction manual'.

Error rate

This is one of the most commonly used measures of efficiency. If a user can complete a task without making any errors along the way, then the task may require less effort than if errors are made and have to be corrected.

Errors can be classified into different types according to why they occur and their severity. A basic distinction is that between 'slips' and 'mistakes'. A *slip* occurs when the user knows how to perform a task, but accidentally does something wrong during the task. An example would be making a typing error when writing your name. A *mistake* occurs when the user has an erroneous model of how a product works. Consider, for example, the user of a computer-based drawing package. Imagine that, in order to create a drawing, the user must first choose from a selection of icons representing various drawing tools (these give the effect of drawing with pencils, pens, paintbrushes, etc.). If, then, a user were to enter the package and then try and create a drawing simply by moving the cursor around the screen this would indicate that they had an incorrect model of how the system worked. It would, then, be a 'mistake'.

Whether an error is a slip or a mistake can have implications for diagnosing usability problems and prescribing appropriate solutions. Slips often occur because the interface to the product is inadequately designed – for example, because a particular button is placed in an inappropriate position or because controls are placed too close together. Mistakes, on the other hand, are more likely to occur because the underlying principles of how a product works are not intuitive or even because they are counter-intuitive.

Whether an error occurs because the user has made a slip or because he or she has made a mistake, the consequences of an error can vary from 'minor' to 'catastrophic'. Here, four categories of error severity are outlined along with their associated costs.

A *'minor'* error is one which the user can notice and correct either instantly or within a fairly short period of time. Returning to the example of the user mis-typing their name, it is likely that the user will be able to spot this straight away and put it right. Thus, it would be only a minor error. Usually minor errors result in a small time cost and only mild annoyance for the user – although if many minor errors occur, these will have a cumulative effect.

A *'major'* error is one which the user is able to spot and rectify, but at a greater cost in terms of time and annoyance. For example, a user trying to record a TV programme with a VCR might notice that they have set the time for recording incorrectly. Although he or she may be able to go back and rectify this, the design of many VCRs would require going through the entire programming sequence again in order to do so. This, then, might be a major error. A question that someone investigating usability needs to address is where the boundary between a minor error and a major error occurs. This should generally be considered in the context of the product under investigation and the expectations of the users with respect to use of the product. For example, the user of a TV remote control might find an error which took 30 seconds to correct to be annoying enough to count as a major error, whilst the user of, say, a complex piece of industrial machinery may regard errors with 30 second time costs as only being minor.

'Fatal' errors are those which prevent the user from completing the task they were attempting. This is because the erroneous action leaves the product in a state that the user doesn't know how to recover from. For example, if the user of a computer-based application program were to quit from the program accidentally and was then unable to get the program running again, this would be a fatal error as he or she would now be unable to complete the task. As another example consider a multi-modal product, such as a stereo system. Imagine that the user were trying to play a CD, but that the system was in 'cassette' mode. If the user did not notice this, then any actions taken with the CD controls would be fruitless and thus, in this case, having the product in the wrong mode might cause a fatal error. Because they prevent the task attempted from being successfully completed, fatal errors have consequences for effectiveness as well as efficiency.

'Catastrophic' errors are those which not only prevent the task from being carried out, but also cause other problems. For example, if the user of a computer-based word

processing package were to highlight all of the text in order to change the format, but accidentally deleted it instead, this would not only mean that he or she hadn't successfully completed the intended task, but also that all work completed earlier would be lost. This, then, would be classified as a 'catastrophic' error.

Time on task

Along with error rate this is probably the most widely used measure of usability. Indeed, it could be argued that it is the most meaningful measure of efficiency as often the cost of errors can only really be assessed on the basis of the amount of users' time that they waste – certainly in the case of minor and major errors. Clearly, the more quickly a user can complete a task with a product, the more efficient that product is for the task.

Mental workload

Mental workload is a measure of efficiency that has been widely used in assessing the usability of products where the time in which to carry out tasks is fixed and where error rates are low. This includes, for example, in-vehicle systems, systems in aircraft and control panels for safety critical processes. The higher the level of mental workload when driving a car or operating a nuclear power plant, the greater the likelihood of an error occurring. For example, Jordan and Johnson (1993) report on field trials in which they investigated the usability of an in-car stereo. They compared the mental workload of drivers in two driving sessions – one in which they were just driving and the other where they were driving and using a car stereo at the same time. These comparisons indicated that operating a car stereo whilst driving noticeably increased the level of workload on the driver and that in-car stereo use might thus be seen as a possible impediment to safe driving. Clearly, with safety critical tasks such as driving it would not be safe to devise studies where there is a significant risk that the driver would actually make an error, as the consequences of this could be disastrous. Mental workload is a very useful measure of efficiency in such contexts.

There are a number of different ways in which mental workload can be measured. Jordan and Johnson (1993) used a subjective technique, the Task Load Index (TLX) developed by the National Aeronautics and Space Administration (NASA) (Hart and Staveland, 1988). This required those participating in the study to complete a brief interview asking how much effort they had to expend with respect to a number of different dimensions. These dimensions included physical effort, time pressure and mental effort.

As well as subjective workload measures there are also physiological parameters that can be used as mental workload indicators. McCormick and Sanders (1983) mention sinus arrhythmia (heart rate variability), electroencephalogram (brain rhythm analysis), pupil dilation and body fluid analysis as being amongst the techniques that various analysts have used. Giving the user a 'secondary' task to do is another alternative. For example, the driver of a car could be asked to count backwards whilst driving. If the driving task were to become highly demanding then it is likely that the driver will have little 'spare' mental capacity for the secondary task. Thus someone investigating the mental workload on the driver would be able to tell when the load on the driver was high as he or she would be likely to make mistakes in the counting task at these points.

Satisfaction: the level of comfort that the user feels when using a product and how acceptable the product is to users as a vehicle for achieving their goals

Qualitative attitude analysis

Perhaps the simplest way of investigating whether or not users are satisfied with a product is to ask them to comment on how they feel about the product. This can be done using a questionnaire or interview (see Chapter 5 for information about these and other usability evaluation methods) or perhaps any comments that people make when using a product can simply be noted. Comments can then be analysed to gain an indication as to the level of satisfaction users felt when using the product.

When interviewing users, for example, the interviewer could ask how the users felt when using the product – was it easy to use, did they enjoy using it or was it frustrating or annoying to use. Similarly, the interviewer might ask whether there were any aspects of the product that users particularly liked or disliked.

Quantitative attitude analysis

Whilst qualitative analysis can often give a useful indicator of satisfaction, there can be benefits in quantifying attitudes to a product. For example, if an investigator wished to compare two different products in terms of users' attitudes towards them or wanted to see if a product met some 'benchmark' level of satisfaction, then being able to put a figure on satisfaction could be helpful.

A number of 'standardised' questionnaire and interview-based tools for measuring the satisfaction component of usability have been developed. These involve asking users to give numerical ratings reflecting, for example, how much they enjoyed using a product, or how confident they felt using a product. Again, more information about quantitative rating scales can be found in Chapter 5, along with examples of rating scales designed to measure the satisfaction aspects of usability.

CHAPTER THREE

Principles of Usable Design

In the previous chapter usability was defined and the importance of designing products to be usable was emphasised. The aim of this chapter is to outline the design characteristics associated with usability. Ten principles of usable design are discussed below along with explanations as to why and how each of these principles affects usability.

CONSISTENCY

Designing a product for consistency means that similar tasks should be performed in similar ways. This will mean that as a user gains experience with a product, he or she can generalise from what has been learned when performing one task with the product to help achieve another. In the context of a computer-based word processing package, for example, the steps involved in the task of putting text into bold format might be as follows:

1 Highlight text to be formatted.
2 Open menu 'Format'.
3 Select the command 'Bold'.

Similarly, the steps involved in putting text into italic format might be:

1 Highlight text to be formatted.
2 Open menu 'Format'.
3 Select the command 'Italic'.

In this case, the procedure for formatting text in bold would be consistent with that for formatting text in italics. This is because both require the text to be formatted to be highlighted and both require the user to select a command from the menu headed 'Format'. These are tasks that users are likely to regard as similar – perhaps the user will think of them as 'formatting tasks' – so it is appropriate that the procedure is similar for both. If the command for putting text into italics were to be placed on a different menu, for example a menu headed 'Font' then these tasks would be inconsistent with each other.

   Inconsistencies are likely to lead to errors. In the example above, if the 'Italic' command was not on the 'Format' menu, but the command 'Bold' was, then it might be expected that a user who had learnt how to put text into 'Bold' would go to the wrong menu when trying to invoke the 'Italic' command, i.e. he or she would go to the 'Format' menu and find that the 'Italic' command was not there.

The layout of the controls in cars is a good example of the benefits of consistency. The foot pedals, for example, are always (at least in manual transmission vehicles) arranged so that the clutch is on the left, the brake in the centre and the accelerator on the right. This sort of consistency means that once someone has learned to drive, he or she can transfer this ability from one car to another. If, however, there was not consistency here – that is to say, if the pedals were arranged differently from one car to the next – drivers would have to put a lot of effort into learning to handle each different car they encountered.

Consistency

Designing a product so that similar tasks are done in similar ways.

COMPATIBILITY

Designing for compatibility means ensuring that the way a product works fits in with users' expectations based on their knowledge of the 'outside world'. Like consistency, it is important because people are liable to try to generalise from one situation to another, and thus a design which facilitates generalisation is likely to be more usable than one which doesn't. The concept of compatibility is, then, similar to that of consistency, the difference being that whereas consistency refers to design regularities within a product or across a range of particular product types, compatibility refers to design regularities between a product and outside sources. These 'outside sources' may be other types of product or, indeed, anything in the 'outside world' which affects the way that the user approaches using a particular product. Consider, for example, placement of the 'Save' command on a menu driven statistics package. Imagine that a user of such a package had never used a statistics program before, but was familiar with other menu-based applications, such as word processing and drawing packages. With these applications, the 'Save' command is nearly always placed on a menu headed 'File'. When trying to save work with the statistics package, then, the user is likely to first look for a menu called 'File'. If, indeed, the 'Save' command is on a menu headed 'File', he or she is likely to find it straight away. In this case, then, the design of the package would be compatible with the user's expectations based on experience with other types of software package. If, however, the command were placed on a different menu, this would be incompatible with what the user expected and would probably cause problems.

Another issue affecting compatibility is what is known as 'population stereotypes'. These are assumptions and associations which tend to be made by nearly everybody within a particular culture. In many cultures, for example, the colour red is associated with danger. Thus when designing, for example, a control panel for a safety critical process – such as the operation of a nuclear power station – it would be sensible if any buttons that operators needed to press in an emergency were coloured red. Similarly, green is often associated with giving permission to proceed (e.g. in the case of traffic lights). It would be sensible, then, to colour buttons green if they were associated with the start-up of a process – for example a button to start up a piece of production machinery.

The colour examples quoted above tend to be fairly universal across cultures, however there are some population stereotypes which tend to be more culture specific. In the United States and mainland Europe for example, the population stereotype is that a

Figure 3.1 The use of red for stop and green for go is an internationally held stereotype. Violation of this principle would cause confusion and in many applications would be extremely dangerous.

switch must be flicked up to turn something on, whereas in the United Kingdom the stereotype is that to turn something on switches are flicked down. Where there are these sorts of divisions, it is particularly important that those involved in product creation take into account the population stereotypes associated with the markets to which they are selling. Again, the issue of safety may be paramount here as users may revert to their instincts in an emergency situation. In the case of switches, for example, an American user might instinctively try to find a switch to flick down in order to shut down a machine.

Compatibility

Designing a product so that its method of operation is compatible with users' expectations based on their knowledge of other types of products and the 'outside world'.

CONSIDERATION OF USER RESOURCES

When interacting with a product a user may be using a variety of their resources or 'channels'. For example, when tuning a TV set, the user will be using their *hands* to push a button on the remote control, their *eyes* to check that the picture is good and to read any information on the screen, and their *ears* to check that the sound is properly tuned.

It is important that when using a product none of the user's resources are overloaded – if this happens there are likely to be usability problems. This book is being written on a computer-based word processing package. Using a word processor is a task which places a high level of demand on the visual channel, with gaze moving backwards and forwards between the screen and the keyboard. Also installed on the computer is an electronic mail (e-mail) program. From time to time messages come into the electronic mailbox and whenever this happens a small icon appears at the top of the screen to indicate the message's arrival. Whilst concentrating visually on what is being typed, this icon is probably too small to be noticeable. However, whenever a new message comes in the appearance of the icon is accompanied by a 'beep'. This sound indicates that something has happened and a quick look up makes the icon visible. This, then, is a simple example of how a design can draw on the audio channel when the visual channel is highly loaded.

As another simple, indeed perhaps rather obvious, example, consider the difference between listening to the radio whilst driving as against watching a TV whilst driving. Driving is a visually demanding task. Clearly, the driver must be aware of the car's position in the road as well as potential obstacles, such as other traffic and pedestrians. Although it could be argued that listening to the radio might cause some distractions, there is no question of it putting a direct load on the driver's visual channel. Watching a TV however would obviously cause additional visual loading and is thus bound to cause a significant and dangerous distraction from the driving task.

A 'traditional' product to which the principle of consideration of user resources has been applied is the piano. Because the piano player needs both hands in order to play the tune, foot pedals are provided in order to either dampen or accentuate the sound. This can be done without any need for the player to remove his or her hands from the keyboard. If manually operated levers were required to do this, then the pianist would encounter serious difficulties!

Consideration of user resources

Designing a product so that its method of operation takes into account the demands placed on the users' resources during interaction.

Figure 3.2 The piano is a traditional design which provides a simple example of how considering user resources can lead to effective design solutions.

FEEDBACK

It is important that interfaces provide clear feedback about any actions that the user has taken. This includes feedback to acknowledge the action that the user has carried out with the product and feedback as to the consequences of any action.

An example of the problems that can be associated with lack of feedback comes from a study reported by Jordan and Johnson (1991). This was a study of the suitability of a hand-held remote control as an input device for operating an in-car stereo. Drivers could use this device to, for example, change the volume of the output, choose a track to play on a compact disc, or alter the balance of the sound from the speakers. In the case of changing the track on a compact disc, there would be a delay of a couple of seconds before the selected track started playing – this was simply due to the time taken for the laser in the player to move to the appropriate position on the disc.

This delay caused problems for the users because they were not immediately sure that they had actually made the necessary input in order to perform the task. This could lead to them pressing the appropriate button again or trying to press some other button on the assumption that the first action that they had taken was incorrect. More seriously, it often led to drivers removing their gaze from the road in order to check that the button they had pressed was the correct one.

A simple solution to this problem would be to have audible feedback (such as a 'beep') whenever the button were pressed. This way users would know that they had taken the correct action and could then turn their full attention back to driving whilst waiting for their selected track to start playing.

The above example relates to feedback acknowledging that an action has been taken. It is also important to provide feedback showing the results of an action that a user has taken. Telephone use provides a simple example. After dialling a number, the user will hear some sort of tone indicating what the result of dialling that number has been – usually either a tone which indicates that the phone which has been dialled is ringing or a tone that indicates that the phone dialled is in use (engaged).

It is important that the feedback given is meaningful. In the case of the telephone tones, the feedback that the user receives does not directly mirror what is happening, they are simply sounds whose meaning the user must learn through experience. This is probably adequate with a simple product such as a telephone, but for more complex products – such as computer-based software packages – more representational feedback can be useful. This is an advantage that can be offered by some graphical user interfaces. For example, consider again the case of formatting text with a word processing package. With some packages a change in the text's format might be represented on the screen by a change in the colour of that text. So, for example, text that the user has put into bold lettering might appear in red on the screen, whilst text that has been put into italics might appear in blue. This is at least giving the user feedback as to the result of an action that they have taken, however the meaning of the feedback relies on users learning and remembering that red text will be printed out in bold and blue text in italics. In other packages, however, the format of the text is represented directly – so that bold text actually appears bold on the screen and italicised text appears in italics. This is preferable as the user has no need to learn or remember any colour coding, but can see the result of actions represented directly.

Feedback

Designing a product so that actions taken by the user are acknowledged and a meaningful indication is given about the results of these actions.

ERROR PREVENTION AND RECOVERY

It seems inevitable that users will make errors from time to time when using a product. However, products can be designed so that the possibility of errors occurring is minimised and so that the user can recover quickly and easily from any errors that are made.

As an example of design for quick error recovery consider a computer-based statistics package. Typically with these packages the user will type in rows and columns of numbers representing the values of certain variables, and will then invoke a command in order to perform some calculation on these numbers. Imagine that whilst typing in the rows of numbers the user types a letter 'o' where he or she should have typed a zero. When the user comes to perform a calculation this error would cause problems as the program would not know how to deal with the letter that appeared in the columns of data. Presumably the program would then return some sort of error message, leaving the user to go and find the offending entry (which might be extremely difficult given that the letter 'o' and zero look similar), correct it and then go back to perform the calculation again. A better solution is if the program spots the error as soon as it occurs and alerts the user to the problem there and then. So, were the user to make the error described, a dialogue box could immediately appear signifying that a non-valid input had been made. The user could then quickly correct the error before moving on with the task.

The 'undo' facilities provided on many software packages are also a good example of how a design can make errors quick and easy to recover from. These are also beneficial in encouraging users to take an exploratory attitude to using the packages. After all, if the user tries a command and something unexpected happens, there is the 'insurance' of knowing that the action can be quickly reversed with the 'Undo' command.

As an example of how errors can be prevented from occurring in the first place, consider the sequence of operations that users have to go through when programming a video cassette recorder (VCR). Here users have to input a series of parameters, including the time at which the programme they wish to record starts and ends, the TV channel that it is on, and the date or day of the week that it is on. They then have to activate the timer so that the VCR will record. Now, were the user to forget to enter any of these parameters or were to forget to activate the timer one of two things might happen – the VCR might not record anything at all or it might record on the basis of the parameters that the user had input plus any defaults that were stored in the system. Thus if the user input parameters for the VCR to record from 17.00 hours to 18.00 hours on Wednesday but forgot to set the channel, the VCR might automatically record, say, BBC 1 at the time and day set. This might not, of course, be the channel that the user actually wished to record.

Many VCRs are designed to avoid such errors of omission by taking the user stage by stage through the programming process. Once users have entered the programming mode they are first requested to input the time at which they wish the programme to start, they are then requested to input the finish time, then the channel, and so on until all the information is there. This is preferable to a design in which the user uses separate procedures to input each parameter, because this type of design relys on users remembering the range of parameters which they must set.

Error prevention and re overy

Designing a product so tha t the likelihood of user error is minimised and so that if errors do occur they can be reco ered from quickly and easily.

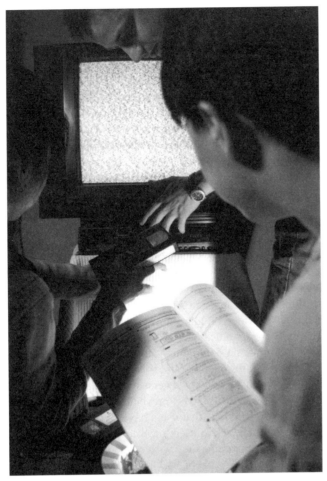

Figure 3.3 VCRs have a (probably deserved) reputation as being 'usability nightmares'. With many designs users have to go through involved sequences in order to program the recording function. One slip can lead to the wrong programme being recorded.

USER CONTROL

Products should be designed in order to give users as much control as possible over the interactions that they have with the product. This means, for example, giving control over the pacing and timing of interactions. A criticism sometimes levelled at speech interfaces is that they take control of the pace of interactions away from the user. For example Jordan (1992b), when considering the use of speech interfaces for in-car systems, notes that sometimes drivers may be presented with information at a time when they are least expecting it. Imagine, for example, that a driver were undertaking some sort of complex manoeuvre, such as moving into a stream of traffic at a roundabout, when suddenly the interface were to give some information – for example that the oil pressure in the engine was too low. It seems likely that the driver would either miss this information because he or she were concentrating on the manoeuvre, or would be distracted from the manoeuvre at a possible risk to safety. A visual display might be more appropriate for non-urgent information such as this as the driver could check this when he or she felt that it was safe and appropriate to do so.

Another type of interface which can remove control from the user are those with time-out facilities in them. Some VCRs, for example, have time-out facilities in their programming modes. This means that if the user doesn't make any inputs for a given period of time (perhaps 30 seconds or so) then the VCR will exit from program mode. An experimental study looking at the usability of VCRs identified this as a usability problem (Jordan, 1992b). Often, when in programming mode, users would stop the programming sequence to consult the manual, only to find that when they returned their gaze to the display, the VCR had tripped back to another mode. Time-out facilities can cause particular problems for users who are new to a product because they may take longer than experienced users to move from one stage of a task to the next. Perhaps a better solution would be simply to include some sort of 'home' button, so that if users felt lost in a system, they could return the system to a familiar mode quickly and easily – with this solution the user would have the choice and would thus be in control.

In the case of computer-based packages this might have implications for, for example, any default settings included in the system. Having default settings can provide benefits to users in terms of the speed with which they can get going with a package. Again, considering word processing packages, imagine the time and effort that would be involved in having to input everything from preferred font size to the width of the margins, before a word could be typed! However, even here it is important that users are aware what defaults have been used and that it is clear how they can be altered should they wish to do so.

Designing for adjustability is another good example of how users can be given control. When designing a chair, for example, the designer should try to ensure that its dimensions are appropriate for the intended users, but at the same time the user should be able to easily adjust dimensions such as the height of the seat from the ground and the angle of the backrest to suit his or her particular preferences.

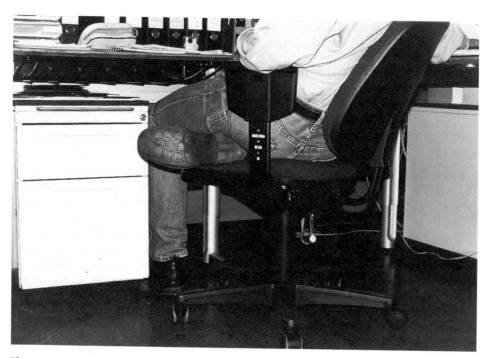

Figure 3.4 Adjustable designs give users additional control as they can tailor the dimensions of the furniture to suit their own needs.

User control

Designing a product so that the extent to which the user has control over the actions taken by the product and the state that the product is in is maximised.

VISUAL CLARITY

It is important that information is displayed in such a way that it can be read quickly and easily and without causing any confusion. This includes both labelling and information displayed as feedback.

Here, those involved in the design of products should take into account issues such as whether or not alphanumeric characters are big enough to read, how much information can be put in a particular space without it becoming too cluttered, how colour coding can be effectively used in an interface (whilst still taking into account that a significant proportion of the population (especially males) suffer from colour blindness), and where information should be displayed.

On-screen interfaces to TV sets are a good example of a product where these issues are important. Because a TV monitor is used, there is the potential for using a lot of colour in the interface. This may be beneficially used for distinguishing between modes – for example controls for altering the sound parameters could be displayed in, say, yellow, whilst those for altering the picture could be in blue. Colour could also be useful where the user must select a command from an on-screen menu. As users scroll through lists the selected command could appear in, say, red.

Those involved in designing on-screen interfaces should also take into account the distance that the user will be from the screen when interacting. Usually these interfaces will be operated via remote controls so that the users will probably be sitting in a chair some distance away when making inputs. Clearly, then, it is important that the characters used in the displays are big enough to read from a typical viewing distance.

Many TVs, especially those at the high end of the market have a plethora of different functions. Those designing the TV might have to consider how many different functions can be displayed at once without causing clutter. The more that can be displayed at one time, the less depth will be required to the menu structure – an advantage as there is less likelihood of the user getting lost in the system. On the other hand, displaying *too* much information at one time can be a disadvantage, because the users will have to search through more information to find what they are looking for.

The issue of where information is displayed is also important. Firstly, those involved in the design of the TV might have to decide whether the whole of the screen should be used as an information display area or just a small part of the screen. Using the whole of the screen is preferable from the point of view of avoiding clutter. It might also allow for bigger textual characters to be used, thus enhancing legibility. On the other hand, the more screen that is filled with the display, the more of the TV picture will be obscured when the user is adjusting the settings. Another question to address as regards where information should be placed is to consider whether or not to put an opaque back-panel behind the menus. This, again, would probably make the display easier to read at the expense of obscuring more of the TV picture. The alternative is to simply use the TV picture as the background to the menus, but this might lead to clutter on the screen depending on what is on the TV at the time.

Of course it is not only on screen-based displays that visual clarity is important. It is also important, for example, in the context of labelling for interfaces based on 'knobs and dials'. So, for example, it is important that the labels for knobs and buttons are clear. With interfaces which contain many dials – for example some control panels – these dials should be clearly distinguishable from each other and should be spaced in such a way that they are in a convenient viewing area, but not so close together as to make the display area cluttered.

Visual clarity

Designing a product so that information displayed can be read quickly and easily without causing confusion.

PRIORITISATION OF FUNCTIONALITY AND INFORMATION

When a product has a vast array of features it may be appropriate to prioritise some features when designing the interface to the product. Prioritisation might be on the basis of, say, how often particular functions are used or the comparative importance of the different functions. Having decided on which functions are most important, those which are considered as having priority can then be given a more prominent place in the design.

The design of graphical interfaces to computer-based applications is an example of where these issues might arise. Often these applications contain hundreds of features which can be invoked via the selection of commands from menus. Whilst the appropriate grouping of commands and sensible and meaningful menu headings will bring benefits in terms of usability, there is still a danger that the sheer number of commands will increase the time that it takes to search for any particular one. A common and effective solution to this problem is to include 'toolbars' in the interface. A toolbar contains icons that represent certain of the product's features. By using these icons the user can activate certain of the commands without having to use the menus. Placing the most commonly used functions on the toolbar can thus save the user a lot of time and effort searching for the commands most frequently used amongst long lists of other less used ones.

Another way of addressing this issue with menu-based systems is to use hierarchical structures. So, for example, when opening a menu, the commands that are most frequently used could be immediately visible, whilst this level of the menu hierarchy could also include 'gateway' commands leading to lower-level menus where the less commonly used commands could be accessed.

The same principle applies to displaying information – just as some functions are more commonly used than others, it is also true that there is some information that users will want to see more often than other information. The use of default display modes can bring benefits here.

Consider, for example, the display screen on a video cassette recorder (VCR). Usually, there is only a small area provided for this, yet, potentially, there is a great deal of information which could be displayed. This includes, for example, the current time, the channel being watched, the mode the VCR is in (e.g. play, record, stop, etc.), and information about any recording sequences that have been programmed in. Clearly, presenting all this information on the display at the same time would cause an immense amount of clutter and would be extremely difficult to read. In order to get around this potential problem, most VCRs are designed such that they have a default display mode – often the current time – with other information accessible via mode buttons. The default display mode represents, in effect, a prioritisation of the information which is displayed in that mode over the other information which could be displayed.

Figure 3.5 Flaps covering some of the less frequently used controls can be a sensible and simple way of prioritising functionality. The controls for the widely used functions are visible at all times whereas those rarely used are hidden away behind the flap. This minimises clutter and makes it easier to perform the basic tasks.

Prioritisation of functionality and information

Designing a product so that the most important functionality and information is easily accessible to the user.

APPROPRIATE TRANSFER OF TECHNOLOGY

Applying technologies that have been developed for one purpose to another area has the potential to bring great benefits to users. However, if done without sufficient thought it can also cause problems.

Consider the history of the TV remote control. This device was originally developed as an aid to disabled people who had difficulty in moving to the TV set itself in order to change the channel, alter the volume, etc. The device was then adopted by manufacturers as something for use by all TV users and now comes as a standard accessory for nearly all TV sets. Indeed, it is now usual that far more functionality can be accessed via the remote than on the control panel on the TV itself. Notwithstanding problems that sometimes occur due to the poor design of individual remotes, this is a good example of how transferring a technology developed for a particular group of users to a wider user population has brought benefits. After all, people would rather stay sitting comfortably in their chairs than having to get up to change the channel or alter some other parameter.

Subsequently, the remote control has also been implemented with other products, such as audio systems, VCRs and lighting systems – again bringing benefits in terms of convenience for the user.

Figure 3.6 Head-up display technology has proved successful in aircraft where the pilot is reading against a background of sky. This technology has proved less successful in road vehicles, where drivers would have to read the display against visually cluttered backgrounds.

Another area in which remote controls have been implemented is for making inputs to in-car stereo systems (Jordan, 1992b). However, the benefits of application in this context are, at best, dubious. A remote control is convenient to use when sitting in front of a TV set as all the user has to do is pick it up, point it at the set and press the appropriate control. These simple actions, however, can become much more difficult when driving a car at the same time. Locating the remote control in the first place can prove difficult. Maybe the driver will have placed it on the dashboard or on the parcel shelf down by the gear lever. Whatever is the case, the driver may have to look around until he or she has spotted it and then, perhaps after having groped around in order to pick it up, will have to orientate it correctly in the hand in order to point it at the stereo. This is, of course, likely to be time-consuming and, more worryingly, will divert the driver's attention from the road. Indeed a study reported by Jordan and Johnson (1991) indicated that using a remote control to operate a car stereo increased the level of demand on the driver compared to the conventional technique of pressing buttons on the stereo itself.

Another example from the in-car environment in which transferring technology might not bring the benefits hoped for is in the use of head-up displays (HUDs). These were originally developed for use in aircraft where they have proved a very successful interface medium. Here, information is projected onto the front windscreen of the aircraft where it can be read without the pilot having to look down at a control panel. This works well because the scene outside the aircraft which forms the background to the information displayed is generally one of clear sky.

If HUDs were to be used in vehicles, however, the view through the windscreen would be much more cluttered – containing, for example, other vehicles, pedestrians, the road, hedges, trees, etc. Any information presented on the screen would have to be read against this background, which may be very difficult. Indeed, presenting information on the screen may simply compound the difficulties of an already cluttered view.

Appropriate transfer of technology

Making appropriate use of technology developed in other contexts to enhance the usability of a product.

EXPLICITNESS

Products should be designed so that it is clear how to operate them. As a simple example, consider the design of doors in public buildings. When someone approaches a door he or she must decide whether it is opened by pushing or by pulling. If the door is well designed it should be clear which is the correct action to take. A flat metallic plate on the door indicates that the door should be pushed whilst a bar that can be grasped indicates that pulling is the appropriate action. This example refers to what Norman (1988) refers to as 'affordances'. Affordances are properties of a design which provide strong clues as to how the product works – in other words they make its method of operation explicit.

With computer-based applications, the representation of commands is an example of where explicitness can make the product more usable. With menu driven systems, commands that are represented explicitly are those where the name of the command clearly signifies its function. For example, the command for downloading the text in a word processing file or the figures in a statistics package onto paper via a printer is usually named 'Print'. For most users, the function of this command is probably clear from its name. If it were the case that this function could only be activated via a function key labelled, say, 'F1' then the representation of the command would not be explicit, as there seems to be no *a priori* reason why 'F1' should represent printing.

Where functions are represented by icons, the design of those icons will also affect the explicitness by which the functions are represented. A study by Maissel (1990) classified icons as to how representative they were – the more representational the more closely users associated the design of the icon with its function. Studies by Moyes and Jordan (Moyes and Jordan, 1993; Jordan and Moyes, 1994) indicated that representationalism had a marked effect on the guessability and some effect on the learnability stages of usability. In other words, during their earliest interactions with a product, users rely on the representational properties of the icons to identify the function that the icon represents. (The representational properties were less salient in terms of their effects on experienced user performance, as by this stage users were often able to remember which icon was associated with which function even if the icons were not representational.)

Explicitness

Designing a product so that cues are given as to its functionality and method of operation.

CHAPTER FOUR

Designing for Usability

In the last chapter the principles of usable design were outlined, along with explanations as to why and how they affect usability. The aim of this chapter is to address the practicalities of creating usable designs by taking a 'user-centred' approach throughout the design process.

SPECIFYING USER CHARACTERISTICS

Before embarking on a design, it is important to be clear about the characteristics of those who will use the finished product. After all, a user-centred approach to design is meaningless without knowing who the users are!

In Chapter 2, the importance of considering user characteristics was highlighted along with examples illustrating that the fit of a product's design with the characteristics of its users is central to designing for usability. So, the first thing to consider is who the product is aimed at. This could be, for example, the general public, a particular section of the consumer population, a small specialist group or even an individual.

The starting point in deciding who the product is aimed at is to consider the brief given by those commissioning the product. This may sometimes include a definite statement about who the target group is. In the case of a consumer product, for example, the designer might be asked to design a product that will appeal to, for instance, both men and women between the ages of 16 and 30 or a product that will appeal to teenagers. In other cases the brief may not be explicit about who the target user group are, so those involved in the design of the product will have to make assumptions about who this would be. For example, if the brief was to design a machine for vending train tickets, it would be reasonable to assume that rail passengers would be the target user group for that product.

Of course, knowing that 'men or women between 16 and 30' and 'rail passengers' were the target user groups for particular products is of little use in itself. What is important is to understand the characteristics of the people in these groups and then to take these characteristics into account when designing a product. So how, having identified the target group for a product, can information be gleaned about the characteristics of that group which can then be used in the design process? Advice is given below on

how to find information about various characteristics of user groups and what the implications of this might be for a design. Note that not all characteristics have significant
implications for all products, however all of the characteristics mentioned below are
likely to be important for at least some products.

Physical characteristics

For many products usable design will be heavily dependent on taking into account the
physical characteristics of the user group. Examples of important characteristics could be,
say, height, reach or strength.

Obviously, the aim should be to design a product such that it is usable by as large
a proportion of the target user group as possible. In the case of products for which
the physical characteristics of users are important, this might mean placing displays
at a height where people can read them, putting controls where people can reach and
making products light enough so that people can lift them. Now consider some examples
relating to these.

Tourist information systems are examples of products with control and display interaction which are intended for use by the general public. Visitors to a town or city can
request information about hotels, for instance, by pressing an appropriate button and the
system will return information on a monitor. The designers of such systems are faced
with having to decide at what height the buttons and the monitor should be placed.
Obviously, if these are placed too high, then shorter members of the public will struggle
to reach the buttons and read the screen. Similarly, if they are placed too low, then taller
users may have to stoop uncomfortably in order to interact with the system. So how
should this problem be approached?

Consider the positioning of the screen for example. Perhaps a reasonable solution
might be to place this at the median eye height of the user group. Once again, however,
the designer must return to the question of who the user group are. Does, for example,
the 'general public' who use tourist information systems include young children? If not
then children need not be taken into account when determining the median height of the
user population. Perhaps it is acceptable to design only for those of, say, 16 and older.
If the system is to be placed in a British town then should it be designed on the basis
of the median heights of the British population or should a world median figure be used
to better cater for foreign tourists who might wish to use the system? What about disabled
users? How can their needs be taken into account?

Clearly, a number of issues have arisen here that it may not be possible to address
without first seeking further advice or information. Perhaps in this case the most useful
next step would be to arrange to meet with a representative of the tourist office to try and
gain clarification with respect to these issues or to observe the use of currently installed
systems to see who the users are and the problems that they have. Techniques for observing users are discussed in Chapter 5.

Reach is an important issue in the layout of controls. In vehicles for example it is
important that drivers can comfortably reach the controls without having to change their
position in the seat or lean awkwardly. Similar considerations are relevant in the design
of control panels.

At first the solution to any potential problems might appear to be to simply place all
the controls as close to the operator as possible. However, especially where there are
many controls, this approach can cause clutter. Having too many controls in one small

area may be awkward for the operator both because of the physical 'fiddliness' of the layout and because of the lack of spatial cues that can be used to identify the control. It is important, then, to investigate how much room there is for the controls based on the reach distances of the group who will be using them. In cases such as these, the aim would probably be to place the controls in a position where at least 95% or perhaps 99% of the target user group can reach them comfortably.

As an example of a situation in which the strength of the users may be important consider the design of portable equipment, for example a portable television. Again, at first sight, the way to address this problem may appear simple – design the television so that it is as light as possible. However there will, in reality, be a number of competing considerations. Whilst users may want a portable TV to be easy to carry, they may also want it to have a reasonably large screen and not be too expensive (thus, perhaps, ruling out some 'ultra light electronics' solutions). Again, then, it would be helpful to have information about the strength of the target user group so that the upper limit on the weight of the TV set could be determined. As in the other examples, this would be dependent on the user population at whom the TV was targeted. If the TV was aimed at young men and women then it could be heavier than if aimed at the elderly.

As well as being related to age and gender, physical characteristics also vary with race and nationality. Comparing the USA and Japan, for example, the average American man is 10 cm taller than his Japanese counterpart and the average American woman is 9.5 cm taller than hers (calculated from tables provided by Pheasant (1986)). It is important that those designing for different markets are aware of these differences. It may be, for example, that a car designed with enough leg-room to suit 95% of Japanese passengers, may suit far less Americans.

So, having clarified who the target group for a product is, how do those involved in the design of products find the information that they need about the physical characteristics of this group?

If the product is being designed for a single individual or a small group of individuals it may be best for those involved in the design to take the measurements or administer the strength tests themselves. This might be done if, say, a product was being designed to meet the special needs of a particular disabled person. However, when the target group is bigger, the designer must look to reference materials such as anthropometric tables. Pheasant (1986) is a useful source of such data. These tables give a vast array of information about body dimensions of various populations. As well as listing the median dimensions for the population, they also give dimensions for the smallest and largest 5%.

Cognitive characteristics

As well as the users' physical characteristics, it is also important to take into account their cognitive characteristics when designing a product. These include, for example, any specialist knowledge that the users may have, attitudes the users hold, or any expectations that users are likely to have of a product. Again, these factors are likely to vary according to who the target group for a product are.

Imagine, for example, that the product under development were a software package for computer-aided design (CAD). Here, it might be reasonable to expect that those who would wish to use such a package would have some knowledge of, say, industrial design or engineering drawing. Thus when creating a CAD system it may be acceptable to design the package with only engineers and industrial designers in mind as potential users. This might have implications for some of the command names and terminology used in the system. It might be, for example, that terminology which would only be understood by those with an engineering or design background could be used in such a product without causing the users any difficulty.

Attitudes that users have based on previous experience can affect how they respond to a new product. This is likely to have implications for the satisfaction component of usability. In the context of computers, for example, it may be that many users will assume that software packages which have a graphical user interface are more usable than those that do not. Even if this is not always the case in terms of the effectiveness and efficiency with which the package can be used, the perception of superior ease of use is likely to affect the satisfaction aspect of usability.

Users will often have expectations of how a product will work for a particular task when using it for the first time. These may be based on, for example, experiences with other products, experience of performing different tasks with the same product, or population stereotypes. Again, many of these issues are discussed in Chapter 2, in the context of the effect of user characteristics on usability. However, for the purpose of illustrating how user expectations can affect design decisions, return to the example (given in Chapter 3) of the layout of the foot controls in a car. Obviously, when people get into a new car they bring expectations of how it will work – i.e. how to drive it – based on their experiences of other cars. The convention is that the leftmost pedal is the clutch, the middle one the brake and the rightmost one the accelerator. It is largely because of consistencies such as these between vehicles that drivers can usually move reasonably comfortably from one car to another without serious risk of accident. Imagine the difficulties that would result if a manufacturer decided that they were going to swap the position of the brake and accelerator on their newest model!

Consistency and compatibility (described in detail in Chapter 3) are the vital design principles if the users' knowledge and experience are to be harnessed in making the product easier to use.

As when considering the physical characteristics of users, those concerned with the design of a product are faced with the issue of how they can glean information about the cognitive characteristics of a user group. If the product is being designed for an individual or a small group of users it might be possible to find out about their characteristics by asking each person individually, via an interview or questionnaire for example. Users could be asked to impart information about experience, attitudes, etc. that are relevant in the context of a particular product. If the product is computer-based, for example, this might mean asking about their experience with computers and the various packages that they have used. If, however (as is usual) the product is for use by a wider cross-section of users, the situation is more complex.

Unfortunately, there are no equivalents of the anthropometric tables in which to look up the cognitive characteristics of users. This means that those involved in product creation must either carry out some survey of their own or try to find the information that they need in the human factors literature. The latter approach may not be productive given the diverse and disjointed nature of the literature as regards these issues – conducting a survey is therefore usually the most appropriate approach. Obviously it will not be practical to interview the hundreds, thousands or even millions of people who might make up a potential user group. However, if a sample is selected carefully, it should be possible (at least to some extent) to generalise from the cognitive characteristics of a small group of individuals to the potential user group as a whole. The question then is, on what basis should people be selected for such surveys?

There appears to be a paradox here. The validity of the survey dictates that those surveyed must be representative of the target user population as a whole with respect to cognitive characteristics, yet the very reason for conducting the survey in the first place is that these characteristics are not known! However, the situation is not hopeless – there are certain 'visible' characteristics of people that can be taken into account which it might be expected would have some associations with cognitive characteristics. These might include, say, the person's job and qualifications, and possibly age, gender and nationality, and indeed many other factors, depending on the product concerned. What is required, then, is for the investigator to make a sensible judgement about which cognitive charac-teristics are likely to be important in the context of a particular product and which visible characteristics these are likely to be linked to. He or she can then survey a sample of people that should be representative of the user group.

As an aid to clarifying a complex issue here is an example. Firstly, imagine that the product being designed were a high-end television set and that the target user group was, say, high income professionals – earning £ 30,000 or more per annum – living in Europe. Imagine also that the company designing the TV were based in London and that they decided that time and financial constraints dictated that they would be able to conduct interviews with, say, 10 people.

So, firstly, what are the cognitive characteristics that are likely to be important in the context of using a high-end TV set? Specialist knowledge could be important – those with a real enthusiasm for technology, and TV in particular, may have a greater understanding of the features on a high-end TV than those without such a specialist interest. If a significant proportion of target users had this specialist knowledge then it might be worth drawing on this in the design. Expectations as to how high-end TVs should work might also be an important factor. Expectations are likely to come from previous experience with other high-end TV sets. Attitudes about what to expect from such TVs in terms of features and functionality is another factor likely to have an effect on usability. Again this is likely to be affected by previous experience with high-end TVs. It might also have to do with, for example, the income of the person interviewed. If the cost of the TV represented 10 weeks' wages to someone, he or she might take a more critical attitude towards the product than would someone to whom the cost of the TV represented only a week's work. There may, of course, be other cognitive character-istics that could affect how usable the TV is for a particular person, but for the purposes of this example, consider further the three mentioned. What visible characteristics might be linked to these?

In the case of specialist knowledge, it might be expected that those who had jobs involving the sale of high-end TVs might have a better than average knowledge of the special features that they contain. Thus, those working in the TV retail industry would be a possible source of interviewees should the investigator decide that it was necessary to interview those with specialist knowledge. Also, those who already own a high-end TV set might be expected to have a greater knowledge of these features than would normally be expected. It might be possible to ask retailers for contact details of those who had recently purchased a high-end TV set.

As was mentioned above, expectations about how high-end TV sets should work and attitudes to them are likely to be linked to previous experience with such TVs. So, a visible characteristic associated with these cognitive characteristics is likely to be ownership of a high-end TV sets. Also the annual income of the user is a visible characteristic that is likely to be associated with attitude.

Having considered which visible characteristics might be linked to the important cognitive characteristics in the users of a particular product, the investigator must then consider how the target group is balanced with respect to these features. In this case the target group is high-income professionals and it appears that income could be a factor which might influence attitudes towards a product. Therefore, it would be necessary to make sure those interviewed were in a high income bracket in order that they were representative with respect to the effect of the product on the attitudes of the target user group (this might rule out many of those involved in the retail of TV sets). In terms of experience of having used high-end TVs (which is likely to affect specialist knowledge, expectations and attitude), the investigator would probably be best advised to interview some people who owned or used to own a high-end TV set and some who never had. After all, the manufacturer is likely to be hoping to sell to first-time buyers of high-end TVs as well as those who already have experience of them.

So, on the basis of what has been discussed so far, it would appear that the investigator should conduct a survey of 10 people in a high income bracket, some of who own or have owned a high-end TV and some who haven't (say 5 of each). But what of other external characteristics such as age or gender – need the investigator be concerned about taking these into account when picking a sample for a survey? According to the analysis above, these factors do not necessarily appear to be that important in this case, thus it might not be necessary to be constrained by such considerations when picking a sample for this survey. Nevertheless, if it doesn't present significant problems in recruiting participants, it may be prudent to try and ensure that there is a reasonable mix of older and younger men and women in the sample in case any differences for age and/or gender do emerge.

REQUIREMENTS CAPTURE

Having identified the target users for a product and the physical and cognitive characteristics of these users, it is now possible to move towards deciding on the usability requirements of a product – this is known as requirements capture.

Some requirements can be specified simply on the basis of knowing basic user characteristics such as the physical and cognitive characteristics mentioned above. For instance, it might be possible to decide on the height of controls and displays simply on the basis of the target group's physical dimensions. Similarly, other requirements – such as the terminology used in command names – may be identifiable on the basis of knowing about the users' educational and professional backgrounds.

However, equally often, determining the requirements for a product may involve having a deeper understanding of the users' lifestyles and attitudes. Suppose, for example, that the product under development were a personal stereo. What would those involved with product creation need to know in order to have a good picture of the product requirements?

One issue that would need to be understood would be how the product would fit into the lives of those using it – in other words to establish the context of the product's use. For example, would people be using the product when on the move or only when stationary? If they are using the product on the move then how are they moving? If people might want to use the personal stereo whilst, say, jogging, this would require a different design than if they were only going to use the product whilst stationary (e.g. sitting outside or sitting in a vehicle). For example, if used for jogging, it may be beneficial to have a mechanism by which the stereo could be firmly attached to the user's clothing, whereas a looser attachment might be acceptable if the user was sitting on a train.

There would also be implications for the specification of the stereo's resistance to movement. This issue might be particularly relevant if the stereo played CDs. It would be reasonable to expect that it would be cheaper to design and produce a CD player that could cope with the movement of a train or a bus, than one which could cope with the comparatively severe movements involved in jogging. Clearly, then, specifying that the player could be used when jogging should only be cited as a requirement if this really reflects how people would use the product, otherwise unnecessary costs will be incurred. Equally, of course, if jogging is an important context of use, then the extra costs will have to be accepted as part of ensuring that the product meets its users' needs.

In order to find out about users' lifestyles and the contexts in which they use products, it may be necessary to involve representative users via empirical methods, such as focus groups, interviews or questionnaires (all described in Chapter 5), to enquire directly as to how the product would fit into their lives. For example, users could be asked to 'talk through' what would be a typical day for them or what they did yesterday. They could then be asked at which points they would wish to use the product under consideration and about other tasks that they might be doing at the same time.

So, continuing with the example of the personal stereo, it might be that a user would like to use the personal stereo when, say, reading the paper whilst sitting in the living room with the family. This might lead to requirements relating to, for example, the cable linking the stereo to the headphones (this should not dangle on the paper) and the noise from the headphones (this should not disturb other family members even if the music is being played comparatively loud).

In addition to finding out about the users' lifestyles it may also be important to have an understanding of certain attitudes that they have which may affect the requirements of a product. Representative users can be asked about their attitudes via interviews and other empirical techniques. Again, staying with the personal stereo example, it is possible that a user might have the attitude, 'stereo equipment should be black' – if there was a consensus on this issue amongst the product's target group, then a requirement might be that the stereo be black.

Other attitudes might be more general – they may not relate to the product specifically but may nevertheless have a bearing on product requirements. For example, if there was a great deal of concern about the environment amongst the target user group, then a user requirement might be that the product be environmentally friendly and that it be seen to be environmentally friendly. Increasingly, usability practitioners are becoming aware of the necessity of understanding the values of the users. Usability, then, is not simply about the effectiveness and efficiency of product use, but also about emotional and hedonic benefits of product use. More detailed discussions of the emotional reaction to product use can be found in papers by Rijken and Mulder (1996) and Jordan and Servaes (1995).

Figure 4.1 In order to have an understanding of the requirements for a product design it is important to understand its context of use. For instance, this athlete is using his personal stereo whilst training in the gym.

USABILITY SPECIFICATION

Having decided on the product requirements these must now be translated into formal usability specifications. Clearly some, such as 'the product must be black' are fairly straightforward, but others will be more difficult. For example, if a requirement were that users should be able to perform various tasks effectively and efficiently with a product, what level of performance could be regarded as effective and efficient and how could the product be checked to ensure that it met the specification?

It is possible to lay down quantitative usability criteria that can be tested empirically. For example, if the product under development were a TV set, then a criterion might be, for example, that '. . . 90% of users should be able to tune in 10 TV channels within 5 minutes of switching on the set for the first time'. This could be tested by '. . . a laboratory-based test involving 20 representative users.' This still leaves the question, however, of the basis on which these specific criteria are set.

Where a product is commissioned by an outside customer, this may already be included in the product specification. For example, if the product were a piece of industrial machinery used for manufacturing components the customer might specify criteria of the nature of '. . . 95% of the workforce must be able to produce 25 components per hour with the machinery and 99% of these components must be within acceptable quality tolerances'. Here, then, the criteria are likely to be based on a financial calculation as to how much performance will have to improve with the new machinery in order to justify investing in it in the first place.

Another possibility might be to take the approach that improvements in usability are required over the products currently on the market. So, considering TVs as an example again, if usability evaluations had shown that with competitors' products 90% of users could tune in 10 channels in under 10 minutes, then a TV interface that enabled 90% of users to install 10 channels within 5 minutes might be regarded as being 'twice as usable' in this respect.

Sometimes usability specifications are dictated by safety considerations. For example, in-car products such as car stereos could potentially distract the driver if they were too demanding to use. Because it is potentially dangerous for a driver to remove his or her gaze from the road it could be specified that, for example, for simple tasks such as altering the volume '... 95% of users must be able to complete the task successfully within 5 seconds without removing their gaze from the road'. With some products where safety is an issue there may be legislation which includes usability criteria that the product must meet. For example, certain standards of legibility are required for warning labels.

If there do not appear to be any firm *a priori* reasons for setting usability specifications at a particular level, then the most sensible approach may be to simply ask users what they would accept as usability criteria. Again, methods such as focus groups, interviews and questionnaires could be used for this. So, for example, potential users of a new word processing package could be asked what they would consider an acceptable time period of usage before they had learnt how to accomplish, say, 80% of basic text formatting tasks, or what they would regard as good performance in terms of the amount of time it would take to find a new command on the menus.

In many product creation environments setting usability criteria can be central to ensuring that those managing the product creation process give the issue of usability serious consideration. To some of these people usability may be seen as an abstract concept which cannot be 'pinned-down' or as something which can simply be left to the judgement of designers or engineers. Setting criteria is a way of demonstrating that user performance with a product is something which can be measured and as something by which product quality can be judged, just as, for example, technical performance is a quality indicator.

ITERATIVE DESIGN AND PROTOTYPING

Having decided on the requirements for the product and set the usability specification, it should now be possible to begin to formulate some product concepts. The principle of 'iterative design' is that product concepts evolve through a design/evaluation cycle. This means that the first design idea will be evaluated in terms of how well it meets the users' needs. The strengths and weaknesses will be assessed and a further design developed on the basis of this. This in turn is evaluated and the process repeated until a product is created which conforms to the usability specification.

There are a number of different prototyping options, of differing degrees of realism and sophistication, which can be used in the design/evaluation cycle. Some examples, along with the phase of the product creation process at which they are most useful, are given below.

Product specification

Perhaps the most basic form of prototyping is to simply give a verbal or written description of the form and functionality of the proposed product. These descriptions can then be discussed with users or checked against particular criteria (derived from the requirements capture) to check whether or not they are suitable. These descriptions might, for example, include a list of the features contained in the product, the technical specifications of the product and details of form such as size, shape and colour. This is often a useful starting point for introducing users to new types of product. For example, over the next few years the expansion of the 'information superhighway' will offer numerous possibilities for bringing new forms of services and entertainment into people's homes.

This level of descriptive prototyping is very useful in these cases as it provides the opportunity to come up with a number of concepts – for example music or movies on demand or home shopping – and ideas about how these concepts might work, without having to invest time and effort creating visual representations of the product that would provide such services. Discussion of the basic ideas might lead to initial concept sketches.

Visual prototypes

These are simply visual representations of a product. They could be paper-based sketches or drawings or on-screen representations created using drawing or computer-aided design packages. They may be supplemented with written or verbal descriptions of a product's functionality or procedures for operation. Initial concepts can be shown to users in this form in order to obtain feedback about the merits of particular design directions in terms of issues such as aesthetics and functionality. It may also be meaningful to ask users about their perception of the ease of use of the concepts illustrated. Clearly, however, users could not really interact with the product illustrated, so making valid judgements about the usability of particular designs might be difficult.

Paper-based simulations may be most effective in situations where user acceptance of a product is likely to be highly dependent on its form and functionality rather than the ease or difficulty of interaction. Consider, for example, the design of household lights and lamps. In this case, user satisfaction with the product may be highly dependent on the product's form, rather than on how easy the product is to operate. After all, lights and lamps can generally be operated without too much difficulty as they tend to have few functions – usually merely an on/off switch or perhaps an additional dimmer function.

Models

Sometimes physical representations, or models, of the product are created. Typically, these will be made of wood or of polystyrene foam. Sometimes they will include additional materials to weight the model so that its weight is accurately representative of the proposed product. Models can be particularly useful in assessing how a proposed product would fit into its environment of use, as well as in checking whether its physical dimensions are suitable for the product's purpose.

For example, the handset of a telephone must be of suitable dimensions and weight for users to be able to hold it comfortably. It would be possible to 'mock up' a model of a proposed handset for users to try before going on to manufacture the proposed telephone.

Screen-based interactive prototypes

These are screen-based representations of products. They offer simulated interactions – for example, by clicking the mouse on a representation of one of the product's controls.

The screen-based representation will then change state to represent how the product would react to a particular interaction.

This type of prototype is particularly useful when there are firm ideas for the form of the product and potential interaction styles, but sufficient uncertainty remains to make it worth checking the interface before going on to build a fully working version of the product itself. For example, it might be possible to create an on-screen representation of, say, an audio system, to check whether users could work out what each of the controls was for and whether information presented on displays was meaningful for them. It may still be comparatively cheap to make alterations at this stage and a number of possible options could be tried before deciding on which one to pursue to manufacture.

Fully working prototypes

These are prototypes which – at least from the point of view of the user – are barely distinguishable from a fully manufactured product. In the case of software products these will be the same as screen-based interactive prototypes – software that supports interaction and responds in the same way that the final product would do. However, where the product has physical dimensions then the prototype will have the same dimensions.

In the case of a stereo system, for example, a fully working prototype would be one where the experience of use was no different to that of using a fully manufactured stereo. This would probably mean that the technological workings of the product would be the same as the final manufactured version – however this does not necessarily have to be so.

One way around having to fully implement the technology at this stage is to develop 'Wizard of Oz' prototypes. These are prototypes that appear to the user to be working as they would if manufactured, but in which the responses of the prototype are actually being manipulated by the investigator or an associate. In the film 'The Wizard of Oz' Dorothy and her friends (the Lion, the Scarecrow and the Tin Man) meet the 'wizard' who comes across as a huge man with special powers and a deep booming voice. However, it turns out that what they have really been confronted with is simply a machine behind which a man is standing pulling levers. Similarly, with Wizard of Oz prototypes the users are unaware of the investigator's influence and think that the machine is responding to their inputs independently.

This type of prototyping tends to be most effective with software-based products. Consider, for example, an 'intelligent' information database such as an expert system. In an evaluation situation users might assume that they were interacting with a PC-based software package. In reality, however, they may be connected to another terminal and the 'computer's' responses to their typed inputs might actually have been typed by an associate of the investigator. However, as far as the users are concerned the responses are coming from the computer and they can rate the product on this basis. Some examples of the use of Wizard of Oz prototyping in product evaluation can be found in papers by Beagley (1996) and Vermeeren (1996).

Iterative design and evaluation is fundamental to creating usable products through a user-centred design strategy. In order to carry out these evaluations there are a number of different techniques available. These will be discussed in detail in the next chapter.

CHAPTER FIVE

Methods for Usability Evaluation

In the last chapter an approach to designing for usability was described and the benefits of an iterative design/test approach were outlined. In this chapter a series of methods are described. The basic structure of each method is given, along with hints as to when it is most beneficial to use each of the methods and the level to which a design must be developed before the method can be applied. Each method has a series of properties which gives that method certain advantages or disadvantages. These include, for example, the time, effort and level of skill and knowledge required to use the method, the facilities and equipment needed to run the method effectively and the number of participants needed in order to gather useful information.

In some cases no participants are required at all – the investigator simply gives an expert opinion or some structured checks are made. These are non-empirical methods. However, most of the methods do involve participants. These are known as empirical methods. In this author's experience there is no substitute for seeing people trying to use a product. Although following the principles of sound ergonomic design will normally bring great benefits to the users there are often cases where users will struggle unexpectedly. This is where the methods involving participants have an added value – uncovering unexpected usability problems. Similarly, users may actually be able to cope easily with aspects of a product where, according to conventional human factors wisdom, they might be expected to struggle. Again, this can only be discovered by involving users or potential users in the evaluation. In this sense there is an inherent advantage in empirical methods, although, as described below, there are some circumstances in which it is not practical to involve participants (for example if confidentiality is an issue or where it is very difficult to find appropriate participants).

Of the methods described below, property checklists, task analyses, expert appraisals and cognitive walkthroughs are non-empirical methods; the rest are empirical methods.

Many of the methods described have their roots in psychology (for example, experiments, questionnaires, interviews, incident diaries), some have been adapted from other disciplines (e.g. focus group, workshops, valuation from marketing), whilst others have been developed specifically for usability evaluation (e.g. co-discovery, cognitive walkthroughs and logging).

The aim of this chapter is not only to describe the methods but also to give a flavour of what it is like to use the methods and advice as to how they can be implemented most effectively. The empirical methods are presented first. The order in which these are presented is based loosely on how tightly controlled the methods are. The section starts with private camera conversations, a method where – save for a briefing from the invest-igator at the beginning – the proceedings are left entirely in the hands of the participant. This is in sharp contrast to the last of the methods presented – controlled experiments. With these the investigator usually keeps a fairly rigid grip on proceedings by requesting the participant to follow a predetermined task by task protocol.

The non-empirical methods are then presented. These are not presented in any particu-lar order.

EMPIRICAL METHODS

Private camera conversations

This method involves participants entering a private booth and talking to a video camera about a pre-defined topic set by the investigator. Participants might be asked to talk about, for example, the way in which they use a particular product, how easy or difficult a product is to use, or how a product fits into their way of life.

Usually the brief given to participants will be of a very general nature. For example, '. . . talk about the context in which you use your stereo'. Such general questions give the participants the opportunity to raise the issues that are important to them, rather than having to respond to a series of more specific questions covering the issues that the investigator regards as important.

A variant on the method is to have two people in the booth speaking to the camera at the same time. This can have two potential advantages. Firstly, the participants can prompt each other by picking up on points that the other has made. For example, if one participant were to mention a difficulty that had occurred when using a product then this might help the other participant to recall similar difficulties that he or she may have had. Another advantage is that the participants might find it easier to talk with another person present, rather than simply talking to a video camera, where, of course, they will receive no direct response or feedback about what they are saying. A disadvantage, however, of having another person present, is that there may be interaction effects between the par-ticipants. This might lead to one participant dominating the session, whilst the other gets little opportunity to speak, or one participant effectively setting the agenda for discussion. It could also be that the presence of another person will inhibit participants in terms of expressing themselves as freely as they might otherwise.

Advantages

Because the investigator is not present whilst the participant talks to the camera, this should minimise any potential investigator/participant interaction effects. This may mean that the participants will be less restrained in their comments than in a situation where they were speaking directly to the interviewer. If the investigator meets the participants beforehand, then there may still be some effect from this, however it should not be so great as it would be if the investigator and participant were face to face for the whole session.

Many participants find the private camera conversation sessions enjoyable to take part in. The atmosphere is, perhaps, a little less formal than with some other evaluation methods and the idea of having the chance to be recorded on video can be appealing to some. This can be good for the public relations of the organisation conducting the evaluation. The method was developed by de Vries, Hartevelt and Oosterholt (1996) who report a positive response on using the method at an exhibition centre and in a high school. Because people enjoy taking part it is also comparatively easy to recruit evaluation participants.

The video tapes themselves can make good 'evidence' when reporting back to a commissioner about the outcomes of the evaluation. Having no investigator present can also be helpful in this sense as there is little room for debate about whether the participant is being 'led' in any way.

Disadvantages

The downside of not having any participant/investigator interaction during the session is that the investigator cannot control the direction in which the session goes. Thus, if the participant's monologue starts to go in a direction which is not relevant in the context of the evaluation, then there is nothing that the investigator can do to move it back onto the important issues.

Because there is likely to be little structure to each participant's monologue, and certainly very little structure across the monologues as a whole, analysis of the sessions can be both complex and time-consuming. Interpretation of certain of the participants' statements can also be difficult because when the tapes are analysed it is too late to question the participant as to what he or she meant when something seems ambiguous.

Co-discovery

This method (described by Kemp and van Gelderen, 1996) involves two participants working together to explore a product and/or to discover how particular tasks are done. The idea is that by analysing the participants' verbalisations the investigator can gain an understanding of the usability issues associated with the product. Usually the participants are friends or at least acquaintances. This is beneficial, as if they know each other they are less likely to feel inhibited in speaking to each other about what they are doing and about their opinions of the product.

The investigator may sit with the participants when they are using the product – perhaps giving instructions or helping whilst they use the product, or maybe asking them questions about what they are doing and thinking. Alternatively, the investigator may simply issue the participants with instructions beforehand and then retire to an observation room to monitor the session or, alternatively, record the session on video whilst he or she is absent. The instructions might be of a general nature, such as to explore the product under investigation, or may request that participants complete particular tasks.

As an example, consider two people participating in a co-discovery session to investigate the usability of an audio system. The investigator might ask them to first explore the system and then to do some specific tasks, such as playing a cassette tape, or finding a particular track on a compact disc. The investigator can observe which aspects of the functionality the participants initially try and, through their verbalisations, find out why. If, for example, the first thing a pair of participants did was to start adjusting the radio tuner, the investigator's interpretation of this action might be very much dependent on the accompanying verbalisations. If one participant had said to the other, '. . . let's try and pick up Radio One', the investigator might conclude that using the radio was a high priority task for users. However, if the verbalisation had been '. . . let's turn the volume up', then the conclusion might be that the tuning knob looked like a volume control and that there was, thus, a usability problem.

Advantages

As the example above illustrates, the verbalisations given during a co-discovery session can help to clarify the meaning of incidents that might otherwise seem ambiguous. This can also be true of think aloud protocols (see later in this chapter), however with these the participant is speaking to the investigator directly and thus may be more inclined to 'rationalise' what they say because the evaluation set-up may seem more formal. For example if a participant were asked to explore a computer-based application, he or she may adopt a more structured exploration technique than would otherwise be the case. This would mean that, when explaining to the investigator what is happening, the participant's actions might seem more 'rational' than they would otherwise be. When speaking with a friend or acquaintance however, participants may feel less pressure to come across in this way and may, thus, approach the exploration in a more natural manner.

As with the think aloud protocol, when participants encounter problems with a product, their verbalisations can give a clear indication of why these problems occur. This can lead directly to diagnoses of usability problems, from which it should be possible to move towards a prescriptive solution.

Video recordings of co-discovery sessions can provide convincing material to show to those using the results of a usability evaluation. Because the conversation is between participants and thus perhaps more spontaneous than if the participant were talking to the investigator, this may convince those interested in the outcomes of the evaluation that what is being said is unprompted and represents the real concerns of the users.

Disadvantages

As with the think aloud protocol, giving verbalisations may distract participants from the task or exploration that they are undertaking. This might mean that any performance data gathered from the study is unreliable. Certainly, it seems unlikely that meaningful data about task completion times could be gathered. It should, however, be possible to obtain meaningful basic performance measures, such as whether or not each set task was successfully completed.

Because this method is one in which the participants have a large degree of control over the topics covered in their discussions, it may not always be possible for the investigator to control the direction in which the discussions go. There is, then, no guarantee that all of the issues which the investigator wishes to cover will be raised. Clearly, there is a trade-off that can be made here. If the investigator wishes to be certain that particular issues are raised, he or she could sit with the participants during the evaluation session and ask about these issues. Alternatively, he or she could include a request to cover these issues in the instructions initially given to the participants. However, the more influence the investigator has over how the session proceeds, the less spontaneous it is likely to become – thus one of the major potential advantages of the technique may be compromised.

Focus groups

The focus group is a group of people gathered together to discuss a particular issue. The discussions could cover, for example, users' experiences of using a particular product, their requirements for a new product, information about the contexts in which they carry out particular tasks or usability problems that are associated with using a product.

A focus group consists of a discussion leader and a number of participants. The leader will have an agenda of issues which will form the borders within which the discussion can proceed. This agenda is usually rather loosely structured, as the aim is to allow participants to take the lead in determining the direction in which they wish the discussion to go. This should ensure that the points raised will be those that are of most concern to the participants. In facilitating the focus group, the leader's job is to ensure that all participants have a chance to voice their opinions. It may be, for example, that some of the participants are more vocal than others, and it is important to prevent one or two people dominating the discussions to the exclusion of the others.

It is also usual for the leader to have a set of prompts. These are for use in the event of the discussion 'drying up' due to participants not being able to think of anything useful to say. It is, however, important that the prompts are simply means of triggering more conversation and that they do not lead the participants into giving particular responses. Making a prompt effective can come down to subtleties in the language used. For example, if discussing the usability of, say, a stereo system, it would be inappropriate to prompt with '. . . don't you find that it is difficult to tune the radio?'. Rather, it would be more appropriate to use a prompt such as '. . . how easy or difficult do you find it to tune the radio'. The first prompt is loaded, as the leader has phrased it in such a way that it may give the impression that he or she believes that the radio is difficult to tune and is asking the participants to agree. The second prompt, however, is phrased in a neutral way. The leader's phrasing doesn't give the impression that he or she is expecting a particular answer, but comes across as a genuine query. Such a prompt simply gives participants something concrete to discuss and should serve to restart the conversation. Prompts should, however, be used only when there seems to be a problem in continuing the conversation and not as a means of redirecting a conversation that is in full flow. Even neutrally phrased prompts have the drawback that they can lead participants into discussing issues which might not have been particularly important to them.

As with all techniques involving open-ended questioning, a problem with analysing the discussion in a focus group is in interpreting why a particular issue has not been mentioned. In the example just given, the reason why tuning the radio had not been discussed by the participants until prompted might be that the issue was of no real interest to them, or it might simply be that nobody in the group had thought of the issue until prompted. The enthusiasm with which the conversation proceeds directly after the prompt can be an indicator of this, but if the leader is in any doubt, then the best thing to do is simply to ask how important the issue is after the prompted part of the discussion is over.

When deciding on the number of participants, the investigator has to consider the trade-off between two main factors. The more people who participate in the group, the more chance for participant interaction. Indeed one of the main advantages of focus groups is that one participant's comments can trigger a useful contribution from another participant. Clearly, the more people that participate in the group, the greater the chance of this happening – if there are too few participants then this effect may not be achieved.

The other factor, however, is that of giving all participants a chance to voice their opinions. In this respect, it is more beneficial to have fewer participants, because if people have to wait for too long before getting a chance to talk they may get bored or frustrated. This may lead to them feeling excluded from the proceedings and make them unwilling to contribute. It is difficult, and probably unwise, to make a general statement as to the 'right' number of participants to have. There is a tradition of using focus groups as a means of market research, where 8–12 participants is generally found to be an appropriate size. However, those investigating usability issues have tended to involve less participants – typically 5 or 6 (e.g. O'Donnell, Scobie and Baxter, 1991; Jordan, 1994b).

As with any empirical evaluation method, care must be taken in selecting those who are going to participate in the focus group. Because of the interpersonal dynamics that are involved with this method, it may be tempting to select participants who might be expected to be particularly vocal in expressing their opinions. This, however, puts a bias on the sample and should not be necessary. Rather, it is the job of the leader to make sure that all participants get involved, no matter how reserved they might be. Learning how to be an effective focus group leader is, perhaps, as much of an art as a science and is most effectively learnt through experience and through watching others.

Advantages

Focus groups can be used at any stage of the design process – participants can discuss a concept, a visual or working prototype or experiences of using finished products (Jordan, 1993). Because the method is loosely structured, participants have the opportunity to raise issues which the investigator may not have anticipated would be important. The group dynamics involved can be particularly beneficial here, because an issue raised by one user may stimulate ideas from others. This makes the method particularly suited to the early stages of the design process – it can be particularly helpful in defining the requirements of a product and alerting the designers to potential usability pitfalls that need to be avoided.

Disadvantages

Focus groups are not a particularly good method for obtaining quantitative data. Although it might be possible to gain very basic information such as the number of users who complained of having a particular problem, focus groups do not provide reliable measures of, for example, the time costs associated with particular types of error (asking users questions about this would be open to large sources of error – for example through inaccurate estimates, memory difficulties plus the social factors of the group itself).

Whilst there are potential benefits due to the group dynamics, these may also bring disadvantages. For example, there is a danger that one or two of the members of the group may prove particularly dominant. This may mean that the opinions that are apparently those of the group as a whole may, in fact, simply reflect the opinions of this individual or individuals. Similarly, there may be someone in the group who is particularly retiring. This person may stay quiet during the discussions and, thus, may not get his or her opinions heard. Both of these problems can be addressed by the leader of the focus group. For example, the leader can make a particular point of asking quieter group participants to give their opinions by directly addressing questions to them. Similarly, the leader can ensure that particular individuals do not dominate the conversation by politely acknowledging their contributions and then addressing a question to another group member. Managing the group dynamics is one of the skills that is necessary in a successful focus group leader.

User workshops

A user workshop involves a group of participants gathered together to discuss issues relating to a product's design and usage. Usually, users will get involved in 'designing' a new product. This might mean simply listing their requirements in terms of usability and functionality. It may, however, include becoming involved with designers in sketching out some ideas for possible designs.

User workshops differ from focus groups primarily in that they involve users in a 'hands-on' way, rather than simply asking them to discuss issues. For example, in a study investigating user interfaces for different cultures (Hartevelt and Van Vianen, 1994) two workshops were conducted – one with Japanese TV users and one with TV users from Europe. The participants in each workshop discussed the context in which they used their TVs and also talked about issues connected with the purchasing and installation of TV sets. They were also shown several TV sets in operation and asked for comments about each. Following this they were involved in a co-design session in which they provided design ideas relating to an interface for a new TV.

Advantages

User workshops represent a very direct way of getting users involved right from the start of the design process. Not only are users asked what their requirements are, they can also become involved in translating these into design solutions. Having users work with designers to sketch out parts of a product can be helpful because it exposes designers directly to the people that they are designing for, rather than having the users' requirements communicated to them via a human factors specialist.

Disadvantages

User workshops can be comparatively demanding for those participating in them – both in terms of the time that the workshop takes and the amount of work that being a participant can entail. This may make it difficult to find participants, unless there are people who are particularly motivated to attend and/or who have a large amount of time available. It is also questionable whether it is advisable to get participants so directly involved in the creation of design solutions. After all, they are not designers and it might be unrealistic to expect participants to come up with workable design solutions.

Although direct communication between users and designers was cited as an advantage of user workshops, there may also be disadvantages of this. It may be, for example, that if designers are present, the users participating feel restricted with respect to communicating their ideas. It could be, for example, that the participants might feel embarrassed about saying things that they felt that designers would regard as 'foolish'. This, then, might lead to incomplete or 'toned down' information as to the users' needs and wishes.

Think aloud protocols

This method involves a participant speaking about what they are doing and thinking when using an interface.

Participants may be asked to perform specific tasks with an interface, or they may simply be given the opportunity for free exploration. Where tasks are set, the think aloud session will, to an extent, mirror a controlled experimental set-up (discussed later in this chapter), with the investigator setting the participant tasks in a predetermined order. Free exploration, conversely, involves presenting a product to a user, who is then simply asked to try doing what he or she will with it. Setting participants tasks is helpful in uncovering specific usability faults in a design, whilst think aloud protocols used with free exploration can provide information as to why users use some parts of a product yet ignore others.

During the think aloud session the investigator will usually prompt the participant, in order to encourage him or her to make helpful verbalisations. These prompts may simply be of the general type, for example, 'What are you thinking now?', or they may be more specific, perhaps relating to a particular error that has been made. An example of a more specific prompt might be, 'Why did you press that button?'. Participants' verbalisations may also provide information about the satisfaction component of usability. These can be encouraged by prompts such as, 'How are you feeling now?'.

In order to run a think aloud protocol, there has to be something for participants to interact with. Thus the method is unlikely to be effective at the earliest stages of the design process. To gain major benefits from using the method it will be necessary to have at least an interactive prototype.

Advantages

Participants' verbalisations make it possible to understand not only what problems they have with an interface, but also *why* these problems arise. This means that think aloud protocols can be an excellent source of prescriptive data, which can lead directly to design solutions. Kerr and Jordan (1994), for example, used think aloud protocols as a method for investigating the suitability of functional groupings in a prototype telephone system. Here, participants' verbalisations were useful indicators of the way in which users might see the relationship between functions and thus the way in which functions could most effectively be grouped.

Because think aloud protocol sessions where tasks are set can mirror controlled experimental sessions in their design, it may also be possible to use the session to gather objective performance data, such as task success and number of errors made. However, it may not be possible to reliably collect data with respect to more sensitive performance measures, such as time on task, as having to make verbalisations may slow the participants down.

Think aloud protocols can also be an efficient way of obtaining a lot of information from only a few participants. This is because each participant can provide such rich prescriptive information. Kerr and Jordan (1994), for example, felt that they were able to draw useful conclusions about the functional groupings that they were checking on the basis of involving only two participants in a think aloud protocol.

Disadvantages

A possible disadvantage of think aloud protocols is related to the possible interference between participants' verbalisations and the tasks that they are performing. It could be argued that, in a sense, participants in think aloud protocol are performing two tasks – not only using the product under test, but also trying to verbalise what they are doing whilst using the product. The problem, then, is that this second task may interfere with the first and any difficulties that the user encounters could, possibly, be connected with the distraction caused by having to make verbalisations.

Another potential disadvantage is that because participants are explaining their actions to the investigator, they may feel tempted to 'rationalise' what they do. This could mean that, for example, where a participant's approach to exploring an interface or trying to complete a task was really rather random, he or she might be tempted to give verbalisations that suggested that he or she was taking a fairly logical approach. Indeed, this effect may also work in reverse, with participants becoming 'trapped' by their verbalisations. If, for example, participants give verbalisations that indicate that they are following a particular strategy, they may then feel obliged to continue with this strategy throughout the think aloud session.

The way in which the investigator prompts participants can have an effect on whether these problems occur. In particular, too much prompting can lead to the participant 'making things up' in order to respond. However, this has to be balanced against the risk of prompting too little, which may lead to the data gathered being less rich than could otherwise be the case. Having a feel for what is the right level of prompting, then, is a skill that is central to running an effective think aloud session.

Incident diaries

Incident diaries are mini-questionnaires that are issued to users in order that they can make a note of any problems which they encounter when using a product. Typically, users might be asked to give a written description of the problem that they were having. They might then be asked how they solved it (if at all) and about how troublesome the problem was. The latter issue might be addressed quantitatively, for example by asking users to mark on a Lickert scale. (A Lickert scale is a numbered scale with verbal anchors at either end. In this case, for example, there might be a five point scale, where 5 represents a very troublesome problem and 1 represents a very easily overcome problem. So, if the user felt that the problem was quite troublesome he or she may mark the scale at, say, 3 or 4.)

The method can be used in conjunction with others in which the investigator is present, in order to gain two different perspectives on the problems that users have – the investigator's perspective and the user's perspective. For example, in a comparison of two word processing packages, Jordan (1992a) used incident diaries in conjunction with a controlled experimental set-up. Whenever participants appeared to have a problem with a set task, the investigator made a note of the problem and also asked the participant to note what had happened in an incident diary. This involved users in describing, in their own words, what the problem was. They were also asked to rate how severe they regarded the problems as being and how difficult they expected them to be to solve. Interestingly, the participants' perception of the problems and their severity often differed from that of the investigator. Using the incident diary in this context, then, provided insights into the concerns of the users that could not have been gained solely from investigator observation.

In Jordan's word processor evaluation (Jordan, 1992a), participants had come along specifically to be involved in the evaluation. They were, therefore, quite prepared to take the time required to fill in the diary. However, in most cases, the diaries will be issued to users to fill in without the presence of an investigator. Because of this it is important to ensure that each diary entry will not take too long to complete. After all, when a user has been having a problem with a product he or she is unlikely to be enthusiastic about the prospect of then having to spend a significant amount of time recording what has happened. There is, of course, a trade-off here. Whilst it is important to keep each entry short, it is also important that the user records enough information to make the diaries useful. This means that when designing incident diaries it is vital to have a good idea of the relative importance of the various types of information that could be gathered, so that the vital questions can be included in the diary and the less important ones left out.

Incident diaries are most useful when relatively infrequent problems occur and the investigator cannot be there to observe them. Reliably completed incident diaries can be a useful guide to an interface's usability profile over time. Any task mentioned at all might be considered as one with which there was a guessability problem (i.e. one which the user had trouble with at the first attempt), whilst those that are mentioned repeatedly might be seen as having learnability problems associated with them, or, if the recurrence is particularly persistent, problems associated with experienced user performance.

The method is usually most appropriate for use with finished products that are already on the market, where the diaries are used for recording problems that occur during 'real life' use. The data gathered can then be used for making decisions about new designs or simply for a usability assessment of the current product.

Advantages

The method is cheap in terms of investigator time and effort because, having decided on a set of questions, the diaries can be sent to as many users as necessary. They are also cheap in terms of the facilities needed to administer them – no laboratory, video or audio facilities are required.

The method is one of the most effective for charting the usability of a product with respect to long-term usage.

Disadvantages

Despite the best efforts of an investigator when designing an incident diary, there is still no guarantee that users will complete them every time a problem occurs. Even if this were the case, there is no guarantee that they will be completed accurately. Users may sometimes lack the technical vocabulary necessary to describe a problem which they have encountered. Similarly, their perceptions of what is causing a problem may not always be accurate. The problem, then, is one of the validity of the data they provide – what is recorded in the diaries may not necessarily be a true reflection of what is really going on.

Feature checklists

In its most basic form a feature checklist is a list of a product's functionality. Users are simply asked to mark against the features that they have used. Knowing which features are used and which aren't is helpful as a means of requirements capture when developing products. Extended feature checklists could ask for additional information, for example, the regularity with which a particular feature is used, whether users realised that a particular feature existed or whether users would know how to use a particular feature if they wanted to.

Features and functionality can be listed in a number of ways, for example as semantic descriptions of particular tasks, or, in the case of software packages, as a list of command names. Visual checklists are one way of increasing the validity of users' responses (Edgerton, 1996). Here the layout of the checklist visually mirrors the product in some way. For example, a checklist designed to elicit information about the use of menu-based commands might be laid out in a way which is similar to the menu layout.

Edgerton and Draper (1993) found that checklists offered considerable advantages over open recall in the context of asking respondents to give information about their invocation of commands on a computer-based software package. Sometimes feature checklists can also be used as an alternative to automatic logging devices (see later in this chapter). Automatic logs – which can be used with software-based products or software-based product prototypes to keep a record of all user interactions – can have the disadvantage that they will also pick up accidental actions. This means that the output from a log might give misleading ideas about the pattern of product usage.

Feature checklists primarily give information about the way in which a product is used, rather than how easy it is to use. However, the checklist could be extended to give an idea about the usability of the product's various features. For example, users could be asked not only about whether they have used a feature but also about whether they would know how to use a feature. It might be expected that where a feature was unused but the user knew how to use it, the feature was simply not useful for the user. However, if the user indicated that he or she did not know how to use a feature, then there might be a usability problem associated with this.

Feature checklists are most effective in the context of finished products that have already been in use for some time. The information they provide can be fed into the requirements phase of a new product design.

Advantages

Feature checklists are a cheap method to use, both because they are undemanding of investigators' time and because they require few facilities – no laboratory or video equipment is necessary, for example. They are an effective means of gaining an overview of the way in which a product is used.

Disadvantages

The method does not provide data that can lead directly to measures of usability. Even if users are asked for more information than simply whether or not they use a feature, the investigator will still need to make a number of judgements of interpretation in order to be able to say anything about ease of use of a particular feature. They are, then, more suited to giving a broad overview of product usage, rather than providing rich data about the user's experience of a product.

Logging use

With computer software and some other software-based products it is possible to install automatic logging devices that keep a record of users' interactions with the product – for example, all keystrokes that users have made can be logged, or all the commands selected from menus. The outcome of using such a device is information about the extent to which a user has interacted with a particular aspect of a product, e.g. the number of times a particular command was invoked.

This information then requires interpretation. If parts of a product's functionality have not been used, or have been used very little, there are usually three possible explanations for this. Firstly, it could be that this aspect of functionality is not useful and so users do not bother with it. An alternative, however, is that although the functionality would be useful, it is avoided as it is difficult to use. The third common explanation is that users did not know that the function existed. In the case of computer-based applications, this third explanation may be particularly common for applications with command line interfaces. These may not be particularly inviting from the point of view of encouraging exploratory behaviour. This contrasts with menu-based applications, where scanning the menus will reveal what is available.

Advantages

With automatic logging the investigator can be sure that, notwithstanding a technical problem, all of the users' interactions will be recorded. This contrasts with the feature checklist method, which relies on users' ability to recall the functionality which they have used. Even the most effectively designed feature checklists are unlikely to yield information which is complete in this sense.

Logging can be comparatively cheap in terms of both investigator and participant time. Although participants may be monitored over a comparatively long period, this would usually be in the context of their usual work, rather than in sessions specially set up to monitor them. Some logs can be programmed to do some basic analysis of the data that they have collected – again this saves time for the investigator. For example, instead of simply getting a printout of each individual interaction with the product, the log might also list the number of times each function has been used.

Because logging relies on keeping records of interactions in a product's normal context of use, the method can provide a high degree of 'ecological validity', i.e. the extent to which the evaluation environment mirrors the environment in which the product will be used. In the case of automatic logging, the evaluation environment usually *is* the environment of use.

Disadvantages

One disadvantage of this method is the ambiguity as to how the data gathered should be interpreted. As explained above, if a feature is not being used, it may not be clear whether this is because it is not useful, not usable, or undiscovered. It may be difficult to come to any firm conclusions about these issues on the basis of the log record alone.

It may therefore be necessary to back up logs with another evaluation method, such as an interview asking users why they haven't used a particular function. In this respect automatic logs have a disadvantage compared to feature checklists, as it is possible to design feature checklists so that users can indicate not only whether or not they have used a particular feature, but also whether or not they know it exists and whether they know what it is for.

Another problem with not knowing why a particular feature has or hasn't been used is that accidental function activations may show up on the log and give a misleading impression of how useful a particular function is. This text for example, is being written using a keyboard which has a help key located next to the delete key. The help key has been hit by mistake many times when attempting to hit the delete key – this happens at least once or twice a day. If an investigator were to look at a log of the commands activated, he or she might conclude that because it had been invoked so often the help function was extremely useful and that it should be placed in a prominent position on the keyboard. In reality, however, all of these activations have been accidental. Indeed, from the point of view of this author, the positioning of the help key regularly causes problems and thus constitutes a usability problem. Clearly, there is a large discrepancy here between the potential interpretation from the log and the reality of the situation from the point of view of the user.

Field observation

Field observation involves watching users in the environment in which they would normally use a product. This provides a degree of ecological validity which might be lacking in evaluations conducted in the somewhat sterile environment of a usability laboratory.

Sometimes the investigator will not set any tasks, but will simply let the users get on with what they would do anyway. Sometimes, however, the users might be set tasks of a somewhat general nature. For example, the investigator might ask them to demonstrate what action they would take if a particular situation arose. There are, however, few controls and balances involved when conducting field studies. The idea is to gain an understanding of how the product performs under natural conditions without imposing boundary constraints that would arise with a set evaluation protocol.

It is important, when conducting a field evaluation, that the investigator tries to ensure that the effect of his or her presence is minimal. If the users are aware that they are being watched, they may consciously or subconsciously alter their usual approach to product use. This would compromise the level of ecological validity. Perhaps the most effective way of minimising investigator presence is simply not to let the user know that he or she is being watched. This could be done, for example, by viewing the users from a distance, or by filming them with a hidden camera. However, using such an approach raises ethical questions. Under the ethical standards commonly accepted by those conducting human factors or psychological evaluations, the right of users to be informed as to what is going on is regarded as central. Evaluations would not normally proceed without the users' prior permission. One solution to this, which might be acceptable in some circumstances, would be to inform users afterwards that data or video recordings have been taken and then ask for their permission to use these for analysis purposes.

Analysing data from field studies can be comparatively complex. Before actually observing users it can be difficult to anticipate what the usability issues will be and thus it may be difficult to decide *a priori* on measures of effectiveness and efficiency by which usability can be judged. Similarly, the real context of use of a product can be such that measures of usability which would have given meaningful data in the context of a controlled laboratory environment can prove too insensitive in these situations. For example, whilst time on task measures might give an indication in a laboratory setting of whether or not a particular design change will have an effect on performance, this effect may be largely hidden due to the noise (other factors and distractions that have an effect on performance) that is usually present in a field observation. This does not necessarily mean that the effect isn't there or that it isn't important. It may simply be that if there is too much noise in the data, the effect cannot be picked up using the size of sample that is typically associated with human factors evaluations.

As an illustration of this point, consider the association between heart disease and smoking. This link is well established statistically and accepted by the medical profession. However, it is very unlikely that it would be possible to pick this effect up from looking at, say, a sample of 10 smokers and 10 non-smokers. Samples of thousands are needed to establish such effects. This is because there may be a lot of noise in the data as well as, possibly, confounding factors. There are many factors other than smoking that could cause heart disease, such as diet. It is possible, then, that in such a small sample, differences between the subjects in terms of diet would hide the effects of smoking on heart disease. It may also be that those who smoke are less health-conscious and therefore also take less care with respect to what they eat – this, then, would be a confounding factor. Thus, even if the smokers group did show a higher incidence of heart disease, it might not be possible to attribute it to smoking, because it could be their diets that were causing the problem.

In the context of usability evaluation, other tasks that the users are doing at the same time as using the product under test are one example of noise. For example, if the user was a secretary using a word processor, then he or she might also be answering telephones, dealing with queries, or breaking up his or her word processing tasks in order to, for example, do some filing. Clearly, the effects of these other tasks may introduce too much noise to be able to meaningfully look at the effects of design decisions on, for example, the length of time required to format a memo. Nevertheless, there may still be an effect there. Perhaps if a secretary were to be watched for a whole week he or she may be able to produce, say, two or three more memos with one word processor than with another.

It might be argued that if an effect is not large enough to show up in these circumstances, then it may not be worth bothering about. After all, does it really matter whether the secretary is able to produce a couple of extra memos per week? If the package was causing problems that took hours to solve, then there would be a real problem, but a couple of extra memos – so what?

In this context, this seems a reasonable argument. However, there are other circumstances in which an extra couple of seconds spent on tasks could have very significant consequences. Consider, for example, a control panel in a nuclear power station. If an emergency arose and the operator had to perform an emergency shut-down procedure, then a couple of seconds extra delay in completing this procedure could potentially prove catastrophic. Similarly, in the context of an in-car stereo, if the user has to take his or her eyes off the road for an extra couple of seconds this could also have potentially disastrous consequences. The importance of small performance effects will, then, be situation dependent. If the investigator decides that such effects are important it may be appropriate to supplement – or in some instances replace – the field observation with a controlled experiment (see later) in which such effects could be isolated.

Advantages

The principal advantage of the field observation is that this method is probably the one that comes closest to being an analysis of a product's usability under 'natural' circumstances (of course the presence of the investigator and/or filming facilities are likely to prevent the situation from being 100% natural).

Disadvantages

The complications in data analysis and the possible ethical difficulties are disadvantages, as are the difficulties in picking up comparatively small effects due to noise in the data. Another disadvantage of field observations is that they are usually only carried out on finished products. In this sense they lack the flexibility of, say, questionnaires and interviews, which can be used throughout the design process. It would not usually be meaningful, for example, to consider using a field study to test a concept or an early prototype. However, there have been cases where field observation of performance with interactive 'Wizard of Oz' prototypes has proved beneficial (e.g. Beagley, 1996).

Questionnaires

These are printed lists of questions. Broadly speaking, there are two categories of questionnaire – fixed-response questionnaires and open-ended questionnaires. With fixed-response questionnaires, users are either presented with a number of alternative responses to a question and asked to mark the one which they feel is most appropriate, or they are asked to register on a scale the strength with which they hold an opinion. Consider the following example of a questionnaire item in which users are asked to choose from a selection of responses. In the context of asking about ease of use, a questionnaire might contain the statement: 'This product is easy to use'. Respondents might then be asked to mark a box to indicate their level of agreement or disagreement with this statement. These boxes could be labelled 'Strongly agree', 'Agree', 'Not sure', 'Disagree' and 'Strongly disagree'. Similarly, if asked how often they use a product, respondents might be given the choice of 'Very often', 'Quite often', 'Occasionally', 'Rarely' or 'Never'. With this type of fixed-response questionnaire it is important that the response choices given cover the full range of possible responses and that the wording can be understood by the respondents. It would, for example, be inappropriate if the alternative responses for a question about amount of usage were to be only: 'Very often', 'Rarely' or 'Never' as this would not give someone who regarded themselves as an 'Occasional' user the opportunity to choose a category that they felt they could agree with. Of course, unnecessarily complex language should be avoided. For example, if asking about how easy something was to use, then the word 'easy' would be more understandable than, say, 'elementary'.

Using scales is one way of simplifying the task of using the appropriate semantics, as here it is usual to only use two semantic 'anchors' – one at each end of the scale. There are a number of pre-developed usability questionnaires available which use this system for responses. For example, Jordan and O'Donnell's (1992) 'Index of Interactive Difficulty' (IID) asks respondents to rate a product with respect to the various components of usability. This is done by marking somewhere along a scale with 'Low' at one end and 'High' at the other. The scale is continuous, although for purposes of analysis Jordan and O'Donnell recommend that the analyst consider it as 20 separate segments, quantifying the subject's response according to the segment that is marked. Theoretically, however, the analyst could break the scale down as finely as he or she wanted. The format of the IID is based on the 'Task Load Index' (TLX) mental workload questionnaire developed by the North American Space Agency (NASA) (Hart and Staveland, 1988). This is one of the most widely used means of measuring the mental effort required to complete complex tasks with products.

The 'System Usability Scale' (SUS) (Brooke, 1996) is another example of a question-naire which employs the technique of asking users to mark scales between two semantic anchors. Here, however, there are five distinct scale points to choose from. The question-naire, which was developed for use in the context of computer systems, lists a series of statements which the respondent then has the opportunity to agree or disagree with, for example, 'I felt very confident using this system'. The scales are anchored with 'Strongly disagree' and 'Strongly agree'.

When designing fixed-response questionnaires to collect quantitative data, it is import-ant to pay attention to the issues of 'reliability' and 'validity'. These are complex concepts, both in terms of how they are defined and how they can be measured. In broad terms, reliability is about the repeatability of what the questionnaire measures, whereas validity is concerned with whether or not the questionnaire measures what it is supposed to measure. In the context of using a questionnaire for a usability evaluation, reliability would relate to whether or not a particular respondent would give the same responses if asked to fill in the same questionnaire on two separate occasions. If this were not the case then responses might be more of a reflection on, say, respondents' moods at the time of completing the questionnaire than on the usability of the product that they had been asked to rate. Even given that a questionnaire is reliable, this does not necessarily mean that it will be a valid measure of usability. A usability questionnaire will be valid only if the questions and the responses given really probe the issue of usability. If the questionnaire is poorly designed it could be that responses will reflect some other aspect of the product such as (for example) its aesthetics or its perceived monetary value.

Using pre-prepared questionnaires, such as the IID, TLX or SUS saves the investigator from having to be concerned with reliability and validity, as those who designed the questionnaires have already carried out checks on these. However, whilst pre-prepared questionnaires may provide a good measure of a product's overall usability, there may often be situations where the investigator wishes to design a questionnaire to address issues relating to a specific product. Then, of course, he or she will have to tackle reliability and validity issues.

With open-ended questionnaires, respondents are asked to write their own answers to questions. For example, a question might be 'What are the best aspects of this product?' or 'What tasks do you find most difficult with this product?'. Open-ended questionnaires can be particularly useful in situations where the investigator does not know what the important issues are likely to be with respect to a design's usability. With fixed-response questionnaires the questions have to be framed specifically enough to make the response categories meaningful. With open-ended questionnaires, however, questions can be framed more broadly, enabling the respondents to highlight the issues that they find most relevant.

Generally, open-ended questionnaires are, perhaps, more suitable for the early stages of a design, before the important usability issues have been clearly defined. Indeed, the qualitative data that they provide can play an important part in defining these issues. In contrast the quantitative data which can be obtained via fixed-response questionnaires, can provide a metric by which to judge usability and, thus, fixed-response questionnaires are more commonly used after users have had a chance to use a new product, or at least an interactive prototype.

Advantages

An advantage of questionnaires is that having once designed a questionnaire and checked it for validity and reliability, it can then be copied and issued to as many people as the investigator feels appropriate at little extra cost. Questionnaires can, then, prove a cheap and effective method for gathering data from a large user population. The method is also versatile in that it can be used at any stage of the design process. Questions can be formulated for requirements capture, as well as for investigating users' attitudes to prototypes or finished products. Because the investigator need not be present whilst respondents are filling in the questionnaires, questionnaires can also have the advantage of being free of investigator effects. With an interview, for example, respondents might consciously or subconsciously gear their responses to what they think the investigator wants to hear. The possibility for anonymity afforded by questionnaires can reduce or eliminate these effects.

Disadvantages

Possibly the biggest disadvantage of questionnaires filled in remotely from the presence of the investigator is that only a small proportion of them are completed and returned. The return rate for mailed-out questionnaires is around 25% (Jordan, 1993). The reason why this is a problem is not the low number of questionnaires completed *per se*. After all, if the investigator wanted a sample of 100 respondents, then he or she could simply mail out 400 questionnaires. Rather, the problem lies in the likelihood that the people who actually take the time and effort to complete the questionnaire will be an unrepresentative sample of those in whom the investigator is interested. Those who complete the questionnaire will often be those with comparatively extreme opinions about the issues that are being asked about.

Consider, for example, a scenario whereby a manufacturer of software for use with home computers decided to survey its customers to check their levels of satisfaction with their products. It seems likely that those who would take the trouble to respond would be most likely to be those who had a particularly strong opinion about the software. If the manufacturers were to make the mistake of treating the questionnaires they received back as being representative of the opinions of their user population, they may come to the conclusion that their users were firmly divided into two camps – those who loved using the software and those who hated it. In reality, of course, these respondents are likely to simply represent the extremes of a much broader spectrum.

The problem of low response rates is likely to be exacerbated if the questionnaires issued are particularly long, as this will increase the time and effort needed to respond. Questionnaires which are to be completed remotely should, then, be as short and concise as possible.

One way to solve the problem of low response rates is to invite respondents to complete questionnaires in the presence of the investigator. Clearly, however, this would put demands on the investigator's time, thus negating one of the main benefits of questionnaire use. Another potential disadvantage of the questionnaire is that more care may have to be taken in the formulation of questions than in, say, an interview. This is because, if the questionnaires are to be completed remotely, respondents will not have the opportunity to ask the investigator about anything which is unclear to them. In an interview situation, if there is any ambiguity in the wording of a question or in the meaning of the various response categories, then the interviewee can ask for clarification. With remotely completed questionnaires however, the respondent must make his or her own interpretation of what is meant. Of course, if the questions are not clear, there is a significant possibility that misinterpretations will occur.

Interviews

Here the investigator compiles a series of questions which are then posed directly to participants. There are three broad categories of interview – unstructured, semi-structured and structured.

In an unstructured interview the investigator will ask respondents a series of open-ended questions. This gives the respondents the opportunity to steer the discussion towards the issues which they regard as important, rather than rigidly sticking to an agenda set *a priori* by the investigator. This type of interview may be most appropriate in situations where the investigator has little idea, in advance, of what the issues of concern to the user might be. Suppose that a high-end television set sporting many new features – for example, to do with sound and picture control – was being considered for the market and that the manufacturers wished to gain a feel for which features were likely to bring the greatest benefits to the users and which were spurious. It might be appropriate, in this case, to simply ask participants who had a chance to interact with a prototype general questions such as what their favourite features were, which features they didn't like and why they liked or disliked a feature. There would, then, be little constraint on the sorts of reply that respondents could give.

With a semi-structured interview, the investigator would normally have a clearer idea of what he or she considered to be the relevant issues for an evaluation and thus of the sorts of issues that they might expect respondents to cover when answering questions. Respondents, then, would be a little more constrained as the investigator would try to ensure that certain points were covered by their answers. This is often done by prompting the respondents as they give their answers. Consider again the example of the high-end television. If the investigator was interested in users' responses to the features in general, but especially their response to some features in particular, then a general question could be supplemented with prompts. So, if the manufacturers were particularly interested in, say, how users would respond to a feature such as a sound equaliser, then the investigator might specifically ask users about this feature as part of asking generally about what they thought the best and worst aspects of the product were.

Because of the prompting, semi-structured interviewing techniques can ensure that a central set of issues are covered by each respondent – this provides the opportunity for a more systematic analysis than might be possible with an unstructured interview. At the same time users still have the opportunity to raise issues that are of particular importance to them.

Structured interviews ask the respondents to choose a response from within a pre-set range. This might mean, for example, asking users to rate the usefulness of particular features on a Lickert scale, or asking them to choose a response or responses from a set of categories. Again, considering the case of a high-end television, this could mean marking against items on a list to indicate which features they particularly liked or disliked. The responses from these interviews lend themselves to structured quantitative analysis. However, in order to be able to predetermine the possible response categories, the investigator must have a fairly clear idea of the issues that need investigation.

Advantages

Interviews are a versatile method in that they can be used throughout the design process. As with questionnaires, questions can be formulated which relate to requirements capture as well as for investigating users' attitudes towards prototypes or finished products.

Because the investigator administers the interview directly to the respondents, the likelihood of the respondents misinterpreting the question to which they are replying is less than that associated with a questionnaire. With a questionnaire the respondent has to make an interpretation of the question based purely on what is written down – if this is misinterpreted, he or she may not be able to give a meaningful answer to the question as intended by the investigator. In an interview situation, however, the respondent is free to ask the investigator about anything which he or she is unsure about. Similarly, if the respondent replies in a way that is not meaningful in the context of the question as intended by the investigator, then the investigator can re-phrase the question in a way that the respondent will understand. The interactive nature of an interview, then, can potentially make the data gathered more valid than that which is gathered from questionnaires.

Another way of considering this advantage is to trade off the validity of the data gathered against the time that goes into preparing the evaluation instrument. Given that a certain level of accuracy may be required, it should be possible to achieve this with less preparation effort with an interview than would be necessary with a questionnaire. When preparing a questionnaire, the investigator must be sure that questions are worded without ambiguity and that the nature of the replies required by respondents is clear. With an interview, conversely, it may be possible to compensate for some deficiencies in question formulation by the two-way communication during the interview session itself.

Another advantage that an interview can have over a questionnaire is the lesser extent to which the respondents are self-selecting. With questionnaires, there is often a low return rate. This is a problem because those most likely to return questionnaires may be those with unrepresentative opinions about a certain issue. In the context of usability evaluation, this might be those with particularly strongly negative or, perhaps, strongly positive views about a product's usability. An analysis based on the sample returned might, then, give a distorted view of how users in general would respond to a product.

With interviews there is still a degree to which respondents are self-selecting – after all, those taking part in interviews must be asked for their consent in advance and there is no guarantee that everyone approached will be willing to give up their time for this. However, once someone has agreed to take part, it would be unusual for him or her not to go on and complete the interview session. This is unlike what may happen with questionnaires, where even those originally intending to fill them in may end up leaving them half finished or may never get round to starting them at all.

Disadvantages

The costs of administering a series of interviews are high in comparison with collating information from a similar number of questionnaires. This is because with an interview the investigator will need to be present in order to ask the questions, whilst with a questionnaire, the respondents would normally answer without the investigator present. Where a large number of respondents are required, this can be very expensive in terms of investigator time.

Another disadvantage of having the investigator present is the risk of having the data gathered distorted by an investigator/respondent effect. Whilst data gathered from questionnaires may give an unrepresentatively extreme view of user opinion, the opinions given in interviews could be unrepresentatively moderate. When giving opinions to another person, it is possible that respondents may not want to give vent to particularly strongly held views which they might have felt more comfortable expressing when given the anonymity afforded by a questionnaire. This is because, when interacting with others, there may be a desire to be seen to be 'pleasant' and 'reasonable'. Perhaps respondents will fear that the interviewer might find them unreasonable or unpleasant if the answers they give to questions are too extreme – particularly, of course, if they are very negative.

Valuation method

The valuation method was designed for evaluating the comparative importance to users of incorporating particular features into a product. It involves asking users how much extra they would pay for a product if it were to contain particular features or were to demonstrate a particularly high level of performance with respect to some aspect of its functionality or design. This method can be particularly useful during requirements capture as a way of comparing the potential benefits of different features. This can help in making trade-offs in situations where there are limits on the functionality that can be included.

Jordan and Thomas (1995) originally developed this technique as a means of comparing the importance of three different design issues in the context of a product containing a printer, i.e. print quality, ease of toner cartridge change and eco-friendliness. After a session in which they used such a product, participants were told how much the product would cost in its basic form and were then asked how much extra they would be prepared to pay for it if it were to perform particularly well with respect to each of the issues of concern. Jordan and Thomas considered that the comparative amounts that users would be prepared to pay were indicative of the comparative importance of each issue to the potential users of the product.

Advantages

The technique is quick and easy to administer and yields quantitative data, whereas most techniques used at the requirements capture stage, such as focus groups, questionnaires and interviews, tend to be more suited to gathering qualitative data. Because quantitative data gives a direct measure of preference, it can be easier to make a judgement about the comparative importance of each of the issues than with qualitative data, where the investigator must make interpretations from participants' comments. Quantitative data may also be more convincing for those commissioning evaluations as they may prefer to see numerical data and statistical analyses rather than what they may see as simply the investigator's subjective interpretations of the data.

Another advantage of this technique is that participants' responses are anchored in a comparatively solid and familiar context – that of making a purchase decision. This is something all of us do virtually every day of our adult lives. More traditional techniques for quantifying subjective opinion, such as the marking of Lickert scales, tend to require users to make more abstract or 'philosophical' judgements. Participants are, after all, more likely to have an understanding of the significance of spending, say, an extra £10.00 or $10.00 on a product than they are about whether a feature should be rated as 'Important' or 'Very important' on a scale.

Disadvantages

Although the data provided by this method might provide a useful indication of the comparative importance of the properties of a product, it would be neither wise nor realistic to expect that the data will really reflect the actual amounts that people would be willing to pay. It is quite a different matter to talk about spending a particular amount of money on something than it is to actually spend it. In any case, making predictions about how much people would really pay for something falls into the domain of market research rather than human factors. It is important, therefore, that those relying on the outcomes of such evaluations are encouraged not to take the results at face value, but to see them simply as indicators of comparative importance.

Controlled experiments

An experiment is a formally designed evaluation with comparatively tight controls and balances. The aim is to remove as much noise as possible from the data in order to isolate effects for performance with the product as cleanly as possible.

For example, with an experiment, there will typically be balances on the order in which tasks are set to minimise the possible effects of knowledge transfer between tasks. Consider, for example, using a word processing package for formatting text and for altering fonts. It might be that formatting tasks are done by highlighting the text to be formatted and then selecting the appropriate command from a menu. To alter the fonts it might be that the text whose font is to be altered must be selected and then a command for the font required selected from a menu. There would, then, be a similarity between these two types of task. Both require highlighting text and then selecting commands from a menu.

If users were always to be set formatting tasks first and were asked to alter fonts later in the evaluation, they might be able to guess how to alter the fonts on the basis of their experience of the formatting tasks. Even if they had struggled with the formatting tasks, then the experience gained from completing these tasks may mean that when they subsequently came to alter the fonts this presented no problem. When looking at the data from an evaluation designed without balances for task ordering, this may give the impression that formatting tasks was a difficult task which caused a lot of problems, but that altering fonts was an easy task, which users had little difficulty in completing at the first attempt. This might lead to the conclusion that the package was well designed with respect to altering fonts, yet badly designed for formatting tasks. Of course, this conclusion is almost certainly erroneous. After all, the two tasks are performed in similar ways.

Experimental balances should eliminate any such effects. For example, half the participants could be set the formatting tasks before the font tasks and the other half the font tasks before the formatting tasks. This should mean that for half the participants learning effects from the formatting tasks should be carried to the font tasks and for the other half effects should be carried the other way – thus balancing out. Overall, it is likely that mean performance should be similar for each task.

Experiments are usually carried out under tightly controlled conditions. This means, for example, eliminating distractions which could potentially interfere with task performance. Potential distractions could come from the environment in which the evaluation is conducted – for example, the sound of others having conversations, or movement in the user's field of vision. Distractions could also arise from having to do other tasks at the same time as using the product under evaluation. In order to eliminate these effects, experiments are often conducted in somewhat 'sterile' laboratory environments in accordance with fairly rigid protocols.

Advantages

Because the data gathered from experiments is relatively 'pure', the method is good for picking up comparatively small effects that might not be detectable with other methods, where there is the possibility that they will be swamped by noise or confounded by other effects. This can make the experiment an effective method for investigating specific design options by direct comparison. For example, in the case of a software package, two functional prototypes could be built which differ with respect to some specific aspect – for example, one could have a command line interface and the other a menu-based interface. By applying the correct controls and balances, it should be possible to gather data which validly indicates which tasks can be more effectively carried out using a command line and which it would be better to activate via menus.

Another advantage gained from gathering comparatively pure quantitative data is that it can be used as material for inferential tests for statistical significance. This, then, can give a comparatively unambiguous indication of whether any apparent effects in mean performance levels reflect systematic advantages for one design over another.

Disadvantages

Perhaps the main disadvantage of experiments is that in order to achieve the levels of control and balance necessary to keep the data free of noise and other effects, the environment and circumstances in which experiments are conducted often tends to be somewhat artificial. There is no guarantee, therefore, that effects which appear to be highly significant according to experimental data, will actually prove to be important when the product is used in a real life context. This is a criticism that has often been levelled at usability evaluation in general (e.g. Landauer, 1987), but which could be considered to be a particular criticism with respect to experimental evaluation.

Another possible problem associated with testing in very artificial environments is that the artificiality may affect the way in which the participants interact with the products. It is possible, for example, that they may treat the experimental session as being some sort of examination – the motivation for using the product will, then, be very different to what might be expected if the product were being used in a real life context. Because the product is isolated from its context of use, those participating in experiments will be giving it their undivided attention. In real life use, however, the product may be used as part of a multi-task situation – the users doing other tasks at the same time and interspersing use of this product with the use of others. The product's usability may be very different under these circumstances.

Consider, for example, the evaluation of two in-car stereos under experimental conditions in the laboratory. If stereo A proved more usable than stereo B in a quiet laboratory environment, with the stereos placed on a table, this does not necessarily mean that stereo A would perform better in actual use. When driving, using the car stereo is a secondary task, so the driver is (hopefully!) unlikely to give his or her full attention to stereo use. He or she is also likely to be performing occasional tasks with other in-car systems, such as adjusting the air conditioning or activating the turn signal indicator. Further, the stereo will not usually be in the driver's line of vision. All of these factors, then, may contribute to making performance with the stereos in a real life setting very different to performance in a laboratory environment under experimental conditions.

Figure 5.1 The usability laboratory at Philips Design in The Netherlands. Laboratories provide ideal surroundings for controlled experiments. However, they are open to the criticism of being too far removed from the real environment of product use, i.e. lacking in ecological validity.

NON-EMPIRICAL METHODS

Task analyses

Task analysis techniques break down the methods for performing tasks with a product into a series of steps. Based on this, the techniques can be used to make predictions about how difficult or easy the tasks will be to perform and how much effort is likely to be required. The output from the most basic and simple of task analyses will give a list of the physical steps that the user must perform in order to accomplish a particular task. More complex task analyses, however, will also take into account the cognitive steps involved in a task. The basic measure of task complexity is the number of steps required to complete a task – the fewer of these, the simpler the task is predicted to be.

Consider, for example, the task of cooking a meal using a microwave oven. The physical steps involved in this might be:

1 Put food into microwave.
2 Set power level.
3 Set timer.
4 Press start.
5 Remove food from microwave.

If the cognitive steps were also to be included then the steps might be as follows (cognitive steps in italics):

1 *Decide what food to eat.*
2 Put food into microwave.
3 *Decide power level needed.*
4 Set power level.
5 *Decide how long required for cooking.*
6 Set timer.
7 Press start.
8 *Listen for chime indicating that cooking is complete.*
9 Remove food from microwave.

A number of 'standard' task analysis methods have been developed. Each of these is characterised by its own individual notation. A notation refers to the way in which tasks are broken down into sub-elements and how the measure of a task complexity is derived. Amongst the most commonly used are the Keystroke Model and GOMS (goals, operators, methods and selection rules) (Card, Moran and Newell, 1983). The keystroke model was developed for use in the context of software-based applications and gives a listing of the key presses required in order to achieve a particular task. GOMS also models cognitive aspects of the task.

Some types of task analyses can also be used to investigate whether the interface to a product demonstrates the design properties of consistency and compatibility. Consistency and compatibility are predictors of how easy a new task will be to perform, given that users are already familiar with performing related tasks with the same product or with other similar products. Probably the most commonly used form of this type of analysis is Task Action Grammar (TAG) (Payne and Green, 1986) and other forms derived from this such as Howes and Payne's (1990) notation D-TAG which relates to display-based interfaces.

Advantages

Using task analysis for usability evaluation does not require the involvement of other participants. This can be an advantage in situations where it would be difficult to recruit suitable participants for an evaluation, or where considerations of confidentiality would make the involvement of participants inappropriate.

Task analysis can also be helpful in terms of prescribing potential solutions to usability problems. Having listed the steps required to complete a task, it may be possible to see how the product could be redesigned to reduce the number of steps involved in the task and thus make it simpler. Similarly, with task analyses that look at consistency and compatibility, the investigator can identify aspects of the product's design which cause the inconsistencies and suggest changes to bring these into line with other parts of the same product or with other related products with which the user might be familiar.

Because many task analysis notations are standardised and require the investigator to follow a particular procedure, these evaluation techniques are less likely to be susceptible to investigator bias than more loosely structured non-empirical techniques, such as expert appraisals (see later in this chapter). If two separate investigators were to evaluate the same product using the same task analysis method, the outcomes should be similar. This may not necessarily be the case with expert appraisal, where the subjective opinions of the investigator are likely to have a large influence on any inferences made about product usability.

Disadvantages

With the exception of the analysis techniques which look at consistency and compatibility, most task analysis models assume 'expert' performance with the system. This means that the rules listed will reflect the most efficient way of performing a task and as such represent a theoretical upper bound on a product's usability (i.e. they are only reflecting system potential). This may not, however, be a realistic reflection of how easy or difficult most users would find the product to use – even those with much experience of it (experienced user performance can fall significantly short of system potential). A study by Allen and Scerbo (1983), for example, showed that the performance of a group of experienced users fell well below that predicted by the Keystroke Model in the context of a text editing task.

There are also problems associated with simply counting the number of steps involved in a task and taking this as a measure of task complexity. Whilst having unnecessary steps in task completion may be detrimental to usability, the required effort associated with each step may be at least as important – if not more important – an influence on usability. In the context of, for example, a menu-based word processing package, the difficulty associated with, say, a formatting task will presumably be dependent on whether the required command is aptly named and whether or not it is on the menu where the user would expect to find it. Although this sort of issue could probably be picked up by a D-TAG task analysis, many task analysis methods have no mechanism for taking this into account.

Indeed, it is a possible criticism of task analyses that, if being used as a basis for design recommendations, they will tend to favour solutions that involve minimising the number of steps required to complete a task, rather than minimising the demands associated with each step. Whilst this may be beneficial to those who are experienced with using a product, it may not be so helpful when designing for the less experienced user. As a simple example, consider the use of confirmatory dialogue boxes in computer software packages. These can be a useful safeguard for novice users, as they can provide the opportunity to rethink before undertaking potentially costly actions. For example, if a word processor user were about to delete a section of text which would be too large to reinstate with an undo command, a dialogue box might appear on the screen warning of what the consequences of this action would be. The user would then be asked to confirm whether or not he or she really wanted to do this. Using the simple metric of the number of steps involved in task completion, an interaction design which included a confirmation box would be rated as less usable than one which didn't – after all the confirmatory action represents an extra step. However, having this confirmatory action could prevent users making costly errors. Further, it may well be that having confirmatory dialogues will

encourage exploratory behaviour with an interface. This is because users may be reassured if the consequences of potential actions are made clear before they commit themselves to taking them.

Property checklists

Property checklists list a series of design properties which, according to accepted human factors 'wisdom', will ensure that a product is usable.

Usually, these will state the high-level properties of usable design, such as consistency, compatibility, good feedback, etc. They will then list low-level design issues relating to these – these might be on the level of, say, the heights of characters on a computer screen or on labels on products, or specifying the position of displays and controls. The idea is that the investigator checks the product being evaluated to see whether its design conforms to the properties on the list. Where it does not, usability problems might be expected.

A good example of a property checklist is contained in Ravden and Johnson's (1989) book *Evaluating Usability of Human-Computer Interfaces: a Practical Method*. Originally designed for the evaluation of human-computer interfaces, this checklist has also been used as a basis for the evaluation of other types of product (e.g. Kerr and Jordan, 1995).

Advantages

Again, as with all non-empirical methods, not having participants can be convenient and it preserves confidentiality.

Another advantage of this technique is that it can lead directly to design solutions. Indeed, the criteria against which the product is being judged can be indicative of what the design solutions should be. As a simple example, consider the legibility of textual button labels. If, for example, the labels had to be legible at a distance of 1.5 metres, then the criteria on the checklist might state that characters should be at least, say, 6 mm high. This, then, is not only giving a criteria against which to evaluate, but also a design solution if the criteria isn't met.

Property checklists can be used throughout the design process. Right from the start, the criteria which they list can be used as part of requirements capture and product specification. They can also be used for evaluation of visual and functional prototypes, as well as the evaluation of finished products.

Disadvantages

The validity of an evaluation carried out using a property checklist is dependent on the accuracy of 'expert' judgement. This is the case in two respects. Firstly, it is dependent on the judgement of the person or people who compiled the checklist in the first place. Whilst some of the items on the list may represent design criteria that have been established on the basis of years of human factors research, others may be more speculative – perhaps simply being a reflection of the checklist compiler's own judgement. Returning to the example given above, there has indeed been much research into the size of characters required for legibility at certain distances and there is now more or less a consensus as to what is required. Consider, conversely, a design property such as consistency. Within the human factors community, there is still much controversy about the benefits of consistency and about what consistency means in the context of design (see, for example, Grudin (1989) and Reisner (1990)). It could be, then, that criteria listed on property checklists for judging interface consistency might be less reliable.

The second expert upon whom the validity of a property checklist-based evaluation may be dependent is the investigator. Returning to the example of legibility, it has been stated above that legibility criteria have been fairly well established. Even here, however, there are complications. For example, the font style of the characters may have an effect on their legibility, as may the lighting conditions in the area where the product is being used. The investigator, then, may have to use his or her judgement in order to make a sensible estimate about the likelihood that a label will be legible, rather than rigidly working to what is stated on the checklist.

Another disadvantage of property checklists is that it is not always possible to judge how great an effect on performance any deviation from the listed criteria will have. When people actually come to use products, they may be able to adapt almost effortlessly to cope with some design faults, yet be severely 'tripped up' by others. The danger is, then, that without actually observing people using products, it may be difficult to tell which faults would prove critical. This can cause problems when making redesign recommendations. Because of the limited time and budgets that are usually available to those undertaking such work, it will normally be necessary to have a clear idea of what the priorities for attention are. This is difficult to do without a clear idea of the comparative seriousness of the various usability faults. This difficulty in estimating the comparative seriousness of faults is, perhaps, a problem that could be said to be associated with all non-empirical evaluation methods.

Expert appraisals

Here a product is evaluated on the basis of whether an 'expert' or experts regard it as being designed in such a way that it will be usable. An expert, in this context, is an investigator whose education, professional training and experience make him or her able to make an informed judgement on usability issues with respect to the product under investigation. In the context of, for example, a computer-based application, the investigator might be someone who is an expert in human-computer interaction (HCI) and the type of application program under investigation.

The sorts of issues covered may be similar to those covered by someone using a property checklist, although with expert appraisal the range of issues looked at may be narrower, with the investigator going into more depth. This is because his or her expert knowledge should give the investigator an idea of which the really important issues are in a particular context, as well as an idea of the details that can really make a difference to the usability of a product of a particular type.

Sometimes more than one expert may give an opinion on a product. They may rate the product separately, or work together to give their assessment.

Advantages

As with all non-empirical methods, no participants are needed and expert appraisal is also a good method for providing diagnostic and prescriptive analysis. The prediction of a usability problem will be based on the diagnosis of a particular fault in the interface. The investigator's knowledge of how to design for usability should lead directly to solutions as to how any problems can be solved.

Suppose, for example, that the product under investigation were a software-based information retrieval system. If the system required that the users type in long command strings, the investigator might predict that this might create usability problems due to the users forgetting or mistyping strings. This could lead directly to a number of potential solutions. For example, the investigator could recommend that shorter and more memorable command strings were used or that the system was redesigned as a menu-based interface.

Disadvantages

Again, as with all non-empirical techniques, there is no direct evidence from users that any usability problems which the investigator discovers will actually cause problems. Users can confound expert expectations by adapting to what might seem to be major shortcomings in a product or by being tripped up by a fault that appeared trivial to the investigator. In the absence of empirical data, therefore, the method is totally reliant on the expertise of an individual investigator or of a group of investigators.

Kerr and Jordan (1994) report an evaluation of the usability of a software-based prototype telephone system in which two HCI experts were consulted in order to make predictions about the suitability of functional groupings on the telephone and their likely effects on usability. They found that, of twelve predictions made by the experts, only five were supported by task performance data in a subsequent empirical evaluation. By contrast, eleven out of twelve predictions made on the basis of data gathered from potential users (using a questionnaire-based tool designed by Kerr and Jordan) were supported.

Figure 5.2 Expert appraisals can be completed quickly, without outside participants, and are a useful source of diagnostic and prescriptive information. However, their validity will depend largely on the skill and knowledge of the individual investigator.

Cognitive walkthroughs

The cognitive walkthrough is a form of expert usability evaluation. However, there is a difference between this method and the 'traditional' expert appraisal. In an expert appraisal, the investigator is looking primarily at the design of the product and trying to predict usability on the basis of fit or lack of fit with the principles of design for usability. With the cognitive walkthrough, however, the expert investigator approaches the evaluation from the point of view of a typical user trying to perform a particular task.

The investigator tries to predict whether or not a user would have any difficulties at the various stages of trying to complete the task. This judgement is based on the investigator's assumptions about the effect that the behaviour of the product's interface would have on the users in the light of their cognitive abilities and expectations.

In order to carry out a cognitive walkthrough effectively, therefore, the investigator must have an understanding of the characteristics of those for whom the product was designed. For example, if the product under evaluation were an X-ray machine for medical use, the investigator would need to have an understanding of the sorts of knowledge and skills that he or she could assume that those employed to operate such machines would be likely to have.

As a simple example of a cognitive walkthrough consider the task of tuning the radio to a particular channel on an in-car stereo system. The steps involved might be as follows:

1   Find tuner button.

2   Press tuner button to start automatic search.

3   Listen for channel.

4   When channel tuned in check display for channel information.

5   Repeat first four steps until required channel found.

The investigator would now make a judgement about the demands on the user in going through each of these stages in order to make a decision about whether the stereo was usable for this task. In this case, the stereo user would be in a dual task situation – driving and using the stereo at the same time. Driving, clearly, is a safety critical task and would, then, be regarded as the main task in this context. So the investigator may mainly be concerned about the extent to which use of the stereo is likely to distract the user from driving.

In step 1, then, the investigator might first try to predict whether or not the visual search time required to locate the tuner button would require the driver to remove his or her gaze from the road for a length of time that would be detrimental to the driving task. This would involve a judgement about the length of time required to locate the button and a judgement about how long the driver could safely remove his or her gaze from the road.

Assessment of steps 2, 3 and 4 would require judgements about, respectively: difficulty of reaching the button and the effect of removing a hand from the steering wheel, the demands on the driver's auditory channel and the level of distraction associated with this, and the visual and cognitive demands associated with reading and interpreting the information on the display. To assess the demands associated with step 5, the investigator would have to make a judgement about how often the typical user might cycle through the first four steps.

In many ways a cognitive walkthrough resembles a task analysis – the method for performing a task being broken down into sub-components. With many task analyses, however, the number of sub-components is taken as the basic metric of a task's complexity, whereas with cognitive walkthroughs the difficulty associated with each step is also taken into account.

Advantages

In addition to the advantages associated with not requiring participants, the method is comparatively quick to administer and leads directly to diagnostic and prescriptive information. It is quick because the investigator does not have to run trials or analyse data. It is diagnostic because any judgements about problems are based on the investigator perceiving that aspects of the design do not provide optimal usability. The investigator will, hopefully, be able to draw on the same knowledge that was used to diagnose problems in order to decide on potential solutions to them – hence the method is prescriptive.

An advantage that this method may have over expert appraisals is that with cognitive walkthroughs the assessment is anchored in the comparatively concrete context of performing specific tasks. With expert appraisals, conversely, the investigator is primarily considering whether the product is well designed from a human factors point of view, on the assumption that this will affect how easy or difficult tasks are to perform. It could be argued, then, that the cognitive walkthrough addresses the usability issues in a more direct way. By looking at the difficulty associated with each substage of a task, cognitive walkthroughs also have an advantage over some task analysis techniques which may rely simply on the number of substages as a complexity metric. That simple metric may not give a valid analysis of usability in situations where the substages are not each of equal complexity.

Disadvantages

As with the expert walkthrough, the cognitive walkthrough relies on the judgement of the investigator. If this is not sound, then the outcomes will not be valid. The criteria on which an investigator would be judged as suitable to carry out a cognitive walkthrough would, perhaps, be more stringent than those required to carry out an expert appraisal or a task analysis. The assumption behind an expert appraisal is that the investigator must have a good knowledge of the general principles of human factors in design. When performing a task analysis an investigator is assumed to have mastered the notation that he or she is using and is assumed to be able to break a task down into its constituent parts. With a cognitive walkthrough, however, the investigator is also directly required to be able to make judgements about users' cognitive abilities, knowledge and skill, as well as about the cognitive skills necessary to complete the task.

CHAPTER SIX

Conducting a Usability Evaluation

In Chapters 4 and 5 the importance of evaluation to a user-centred design process was highlighted and a number of methods for evaluation were presented. However, before choosing a method and carrying out a usability evaluation, a number of factors must be taken into account – from the purpose of the evaluation to the constraints and opportunities arising from the circumstances of the evaluation, from deciding on the type of data required to deciding how the evaluation should be reported. These issues are discussed in this chapter. The chapter ends with a case study illustrating the evaluation of a computer-based statistics package.

PURPOSE OF THE EVALUATION

Before embarking on a usability evaluation it is important to be clear as to what the purpose of the evaluation is – this will underpin any decisions about methodology, data gathering and reporting that are made. The purpose of the evaluation will usually be dependent on the stage of the product creation process (PCP) at which the evaluation will take place. Examples are given below of possible evaluation requirements at various stages of the PCP, together with the implications for the type of evaluation to be conducted. In this section implications for the choice of method will be mentioned. Advice as to the specific methods that are most suitable at each stage will be given in the case study later in the chapter.

Benchmarking existing products

Before embarking on the development of a new product it may be beneficial to conduct evaluations of existing products in order to gain an understanding of the state of the art in terms of the usability of what is already available. This might form a baseline against which the usability of the new product can be judged.

Consider, for example, the design of an interface to a high-end television set. Clearly, those developing the set would want to offer customers benefits over what is currently on the market, including benefits in terms of usability. By gathering quantitative data about performance with existing TVs and users' attitudes towards them, it is possible to set a minimum usability specification for the product under development. So if, for example, users were able (on average) to install 10 channels within 5 minutes of first switching on the most usable of the currently existing TVs, then in order to gain a significant usability advantage, it might be specified that on average users should be able to install 10 channels within, say, 3 minutes of first turning on the new TV.

Requirements capture

Early on in the PCP, evaluations of basic prototypes, such as product specifications or perhaps visual prototypes, might be undertaken in order to facilitate requirements capture. Here the emphasis would be on generating ideas or on suggesting some of the characteristics that a product might have. At this very early stage of the PCP it is important not to constrain users in expressing their ideas – the evaluation should therefore be conducted in a way that is conducive to expressiveness and imaginative thought.

A particular point here is that it is often beneficial to free users' thoughts from constraints of technology and perhaps from the opinions of those directly involved in product creation. After all, there is little point in involving users at this stage if they are simply going to be led into confirming the ideas already put forward by the project team.

Evaluating initial concepts

At this stage in the PCP there will typically be several design directions that could still be taken – the aim of a usability evaluation would be to steer the direction of the design in the light of reactions of potential users. Typically, there would be visual prototypes at this stage, either on paper or screen-based. It is also possible that some or all of the functionality would have been implemented on any screen-based prototype, but generally the sort of investment in time that would be required to achieve this would not be made this early in the design process – indeed not until those involved in the PCP were more sure (hopefully on the basis of evidence from usability evaluation!) of which concepts they felt were worth pursuing.

Developing a concept

Having set the design direction, the next stage in the PCP would be to develop this direction into a workable design solution that would satisfy the user requirements. By this stage interactive prototypes would probably have been developed. This means that it may be meaningful to make performance measurements as well as asking users about their subjective reactions to a design. For example, if a software package – say for statistics – were under development, and a decision needed to be taken as to whether an icon or a menu command should be used to invoke a particular function, then two alternative interfaces, one using each, could be evaluated.

Testing a design

By this stage in the PCP much of the design will have been finalised. It should now be possible to build a fully functional prototype which can be evaluated for usability before the product is launched. At this stage, the evaluation could have one or more of several purposes. It might be that the results of the evaluation will contribute to a 'stop/go' decision with respect to release of the product. This is particularly the case where the usability of the product is likely to have implications for safety. For example, if a car stereo proved extremely demanding to operate and distracted drivers' attention from the road, then the manufacturer would (hopefully!) not release it onto the market until it had been made more usable.

Another potential purpose for a test at this stage would be for those involved in the design of the product to see how well they had done in terms of usability and to learn from this, for input in to future designs. The attitude here, then, might be that usability problems probably wouldn't be disastrous (so the product can still be released), but that it would clearly be desirable to minimise these problems in future products. This can often be the motivation for testing when the product under development is a novel one or contains a novel design aspect – in these cases it is often perceived as being vital to be first on the market with a product, but also to use the PCP for this product as an opportunity for knowledge gathering so that the manufacturing organisation builds up expertise in that area. For example, if a manufacturer were the first to develop, say, 'surround-sound' television, then they would want to ensure that they got the product on to the market before their competitors. However, there are particular usability issues which are likely to be raised by such a product – for example, does controlling the multi-directional sound system cause the users serious problems? By testing at this stage issues such as this can be addressed.

Sometimes, even at this late stage in the PCP, it will be possible to make alterations to a product. Often, this will simply be the opportunity to 'tweak' one or two aspects of the design before the product is launched – for example, the colour of the buttons on a telephone or the fonts used to label the keys. In the case of software-based products it may be possible to make comparatively greater alterations than with 'hard' products. However, it would be unusual if an evaluation at this stage were to influence fundamental aspects of the design.

Evaluating a finished product

This refers to evaluating a product that has been released and is in use. If a manufacturer were to evaluate the usability of their product at this stage it would usually be so that they could gain knowledge to feed into their future designs. An advantage of evaluating finished products is that, once they have been on the market for a while, there will be a number of people who have 'real life' experience of having used them. They will be able to report on the positive and negative aspects of using the product in its real context of use – something which it can be difficult to predict from laboratory testing with prototypes.

We have come full circle – essentially evaluating finished products is the same as benchmarking existing products. Evaluation of finished 'Product A' can be used as the benchmarking evaluation for 'Product B'. However, evaluation of finished products can be carried out for reasons other than benchmarking. For example, a potential

buyer might carry out usability tests on a number of products before making a purchasing decision. Consider, for example, a large company making the decision as to
which word processing package they wished to use as their office standard. It would
seem sensible for this company to buy a few packages and carry out some testing
before buying a licence for a particular package to cover all of their sites. Here, then,
the emphasis would be on comparative evaluation – which of the packages is the
more usable. There would, however, be little or no interest in why the packages
differed – after all the evaluators are not interested in designing word processing
packages themselves.

SELECTING EVALUATION PARTICIPANTS

Having decided on the purpose of the evaluation which, as described above, is largely
dependent on the stage of the PCP reached, the next step might be to consider who will
take part in the usability evaluation – some examples of who participates in evaluations
are given below.

No participants

The non-empirical evaluation techniques do not require that anyone participates. Four of
these – task analyses, property checklists, expert appraisals and cognitive walkthroughs –
were outlined in Chapter 5. Sometimes considerations of safety or confidentiality dictate
that no participants should be used for an evaluation. On other occasions difficulties in
obtaining a suitable sample of evaluation participants may make non-empirical evaluation
an attractive option.

Colleagues

Another option when confidentiality is a consideration is to carry out evaluations using
colleagues as participants. This seems to be a common practice in human factors. However, the outcomes of such evaluations can be misleading. The reason for this is because
these people are likely to have a vastly different experience and knowledge of the products that their organisation produces than the typical end-user will. This is likely to result
in unrepresentative levels of performance with a product as well as differing views and
expectations to those of the end-user.

   As a rule, therefore, it is better to avoid involving colleagues as participants. Indeed,
it is often preferable to use a non-empirical method rather than using participants whose
performance and attitude might lead to completely inappropriate conclusions being drawn
about a product's usability. Note that if colleagues must be used, it is best to use those
with as little knowledge and experience of the product under test as possible – for
example, when evaluating reactions to a design, it might be better to ask secretaries rather
than designers. Nevertheless, even being a member of the organisation is likely to affect
attitudes to a product, even if it doesn't affect performance. It would almost certainly be
inappropriate to expect the attitudes expressed by such a sample to generalise to the end-
user population.

Figure 6.1 A secretary participating in the evaluation of one of her company's products. It has been suggested that psychology in the 1960s and 1970s was simply the study of undergraduate psychology students. Thomas (1996) suggests that much usability evaluation is simply a study of the reaction of secretaries to product design!

Representative sample

A far more valid approach in circumstances where confidentiality is less important is to try and find a sample of participants whose characteristics mirror those of the intended end-user population. If, for example, the product is aimed at women aged between 30 and 50, then the participants in the sample should be women between the ages of 30 and 50. A number of user characteristics were mentioned in Chapter 2 and, of course, the investigator should try to match the characteristics of the sample of participants as closely as possible to the characteristics of the end-users.

Because of practical constraints it is almost always the case that the characteristics of the sample can only partially mirror those of the intended end-user population. For example, consider again a target user group of women aged between 30 and 50 and assume also that the target area for the product was Europe and that women of all socio-economic levels were the target group. Assume that the manufacturer of this product was based in Britain. Would it, for example, be good enough to involve 12 women between 30 and 50 registered with a local temporary employment agency as participants? Clearly, it is not possible to give a pat answer to a question such as this. It will depend on the product and what is being investigated in the evaluation.

If, for example, the product was a domestic product, such as a vacuum cleaner, and the evaluation was investigating attitudes towards the product, then it might be expected that women's attitudes towards domestic chores would play a role in the attitude that they had towards the product. In the United Kingdom there are comparatively few full-time housewives compared to other European countries (Economist, 1995). Thus it may be that British women would see domestic chores as being less central to their lives than their European counterparts would. This in turn might lead to different attitudes towards products associated with such chores. Furthermore, because all the women in the sample were registered with temporary employment agencies, they may also be a fairly unrepresentative sample in terms of the split between the time they spend at home and at work.

If, however, the purpose of the evaluation were to look at user performance with the cleaner, then there would be no particular reason to suspect that British women would perform any differently from their European counterparts. After all, the physical and cognitive characteristics of British and European women are probably not so different as to make performance levels with such a product radically different.

Real end-users

In cases where a product has already been released, it may be possible to carry out evaluations with real end-users. Clearly, there will be an advantage in including participants who have experienced the product in a real life context. When constructing evaluations involving participants who have never used the product in a real life situation, it usually falls to the investigator to make some sort of prediction as to the sorts of problems which might occur and the contexts in which the product is likely to be used. The evaluation design will be influenced accordingly and there is a danger of the evaluation context becoming rather artificial. Conversely, by asking 'real' users about their experience with a product and by observing them in action it should be possible to assess how well the product performs when it really counts.

Another advantage of involving real end-users is that these people are more likely to be motivated to use the product than are specially recruited evaluation subjects. This is likely to affect their performance and attitudes to the product. Consider, for example, users of a computer-based word processing package. Imagine that there was a usability fault in the package that caused a problem every couple of days or so – for example, say that the package contained some function for creating a table of contents and that this function tended to cause users difficulty. If an evaluation participant were set a task involving this function in an artificial evaluation, he or she may encounter difficulties with it and perhaps be unable to complete the task set. However, by the end of the evaluation session, he or she may not be too bothered by this and thus when asked to rate the word processing package for usability may take the attitude that most things seemed OK and that, therefore, a high rating is in order.

This might contrast strongly with the attitude of the real life user who may find that the usability problem within this function ends up costing him or her valuable time and effort on each occasion that he or she writes a report. This user might feel so annoyed by this problem that he or she has developed a negative attitude towards the package as a whole, largely on the basis of this particular problem. Here, then, an evaluation involving real users may reveal that this problem is seriously undermining the usability of the product as a whole, whereas a test with recruited participants might have given the impression that the fault was a bit of a nuisance but nothing too serious.

The entire user population

In terms of finding a representative sample of users this is the ideal scenario, i.e. have everyone who uses or is going to use the product involved in the evaluation. Of course, in the case of products that are to be mass marketed, this would almost certainly not be a practical scenario. However, there will be instances where products are manufactured specifically for an individual or a small number of people, where it will be possible to involve all these users in a usability evaluation.

An example of where this can occur is in the design of a product to meet the special needs of a particular disabled person, e.g. in the design of a workstation which enables a disabled person to perform a particular job.

Another example of where the whole user population of a product can become involved in its evaluation is with specialised software packages which may be designed to meet the needs of a particular business. For example, the accounts department of an organisation may have had a package designed specially for their (say) 25 accountants to use. It should then be possible to perform an evaluation involving all of these users.

TYPE OF DATA REQUIRED

Having identified the requirements for an evaluation and considered the people who might be involved, the next stage is to consider the type of data that is required from the evaluation.

Broadly, there are two types of data that can be elicited from usability evaluations – quantitative data and qualitative data. Which is the most appropriate type will depend on the purposes for which the evaluation is being conducted and the circumstances and conditions of the evaluation.

Quantitative data

Quantifying usability will be necessary if usability specifications for a product are to be set and adhered to. Quantitative data can also give a clear indicator by which the effect of particular design decisions can be judged. Quantitative data can be gathered both as measures of performance and as measures of attitude.

Performance data

Performance measures, such as time on task and error rate, are perhaps the 'traditional' metrics by which usability has been most often measured. These measures provide a way of quantifying the 'effectiveness and efficiency' components of usability (see Chapter 2 for a fuller discussion).

Quantitative performance data can be useful in situations where a design decision has to be made and a number of possible solutions are being considered. For example, imagine that the product under development were the remote control to a high-end TV set with a menu interface. There might be, say, two possible remote control designs that were being considered for this – one involving a separate colour-coded button for each menu and another with a cursor key for moving through the menus and an enter key for making a selection. By gathering quantitative performance data with respect to each of these, it should be possible to determine which of the two supports quicker task completion.

Although the example given here related to the design of TV sets, quantitative performance data are probably most commonly associated with professional products where effectiveness and efficiency may be seen as the central usability issues. For example, the output of a factory is likely to be dependent on the usability of its machines – the quicker and more accurately that workers can complete the manufacturing process, the more productive the factory will be and the less waste there will be.

Attitude data

Just as usability specifications can be set in terms of level of performance with a product, it is also possible to include attitude components in these specifications. In order for this to be meaningful, however, it is necessary to quantify attitude, so that reactions to a particular product can be assessed in relation to the criteria set. To talk in terms of '. . . people having positive attitudes to a product . . .' is vague and imprecise. However, to say that the product design should be such that '. . . 85% of respondents give the product a rating of 60 or more on the System Usability Scale . . .' is a concrete criterion that can meaningfully be included in a product specification. Setting such quantitative criteria can also be a way of getting user attitude accepted as a product quality indicator in the same way as technical and performance aspects of a product are. (The System Usability Scale and other instruments for quantifying attitudes are discussed in more detail in Chapter 5.)

Qualitative data

Qualitative data can be useful for a number of reasons. On one level it can be used as an approximation to quantitative data in order to make a 'first pass' at addressing an issue. But, perhaps more importantly, it can provide rich descriptive data which can be used to diagnose usability faults and prescribe design solutions.

Performance data

An example of how qualitative performance data can be useful is in identifying general areas or aspects of a product that may contain usability faults. For example, in the case of the evaluation of a high-end telephone system an initial broad investigation may be conducted in order to gather qualitative information of the sort, '. . . users seem to be having problems with the "follow-me" function'. Subsequently, a quantitative investigation may be directed at this specific part of the interface in order to put a figure on the proportion of users who didn't know how to operate this function, so that the seriousness of the problem could be quantified.

Often qualitative performance data can prove extremely valuable in diagnosing a usability problem and prescribing a solution to it. Suppose, for example, that the manufacturers of a software-based word processing package decided that they were going to release an updated version of the package and wanted user input into the new design from those who had experience of the previous package. The type of qualitative data that it might be possible to glean could be along the lines of '. . . I have trouble putting the text into bold writing as I always go to the menu headed 'Format' when the command that I should be using is actually on the 'Font' menu . . .'. Data such as these are useful in three ways. Firstly, the investigator knows that the user has encountered a usability fault – he or she is having trouble putting text into bold writing. Secondly, the cause of this fault has been diagnosed – the user was expecting to find the appropriate command on a different menu. Thirdly, a solution to this user's problem has been prescribed – move the command to a different menu.

Attitude data

Qualitative attitude data can be useful in ways that are generally parallel to the uses of qualitative performance data. Again, one of these is to direct the investigator towards the particular aspects of a product that appear to be causing dissatisfaction. So, for example, an initial investigation might contain an open-ended question aimed at directing the investigator towards the aspects of a product that a user found particularly irritating. Again, consider that the product under investigation were a high-end telephone system. Responses might indicate, for instance, that users did not like the 'beeping' noises that were made each time a button was pressed. This, then, could direct the investigator to look further at this issue – for example by setting up an evaluation of users responses to various different 'beep' sounds.

Again, like qualitative performance data, qualitative attitude data can provide both diagnostic and prescriptive information. Using the example of a computer-based word processing package, imagine that a user had said something along the lines of '. . . I don't like the command names used in this package, they sound too technical'. Again, this would reveal three things. Firstly, that this user was dissatisfied with the command names used. Secondly, that the problem could be diagnosed as the technical nature of the words and thirdly, that changing the words to those of a less technical nature could be prescribed as the solution.

CONSTRAINTS AND OPPORTUNITIES

Whilst the investigator may have clear ideas of the approach that he or she would ideally like to take in an evaluation, the reality of the circumstances in which evaluations are conducted means that there will almost certainly be constraints on what is practical. Similarly, the circumstances of an evaluation may also provide special opportunities which the investigator can take advantage of.

Time is usually the major constraint on any usability evaluation. However, time can be constrained in a number of different ways, each of which may affect the evaluation differently.

Deadlines

The deadline for an evaluation will affect the time available from the moment that the investigator starts designing the evaluation until the moment that the results of the evaluation must be reported. This, then, is likely to have an effect on the whole evaluation process.

If the deadline is short, the investigator will either have to choose a method that is not too costly in terms of time or to 'cut corners' in order to save time. The latter approach is known as conducting a 'quick and dirty' evaluation. Bonner and Cadogan (1991) refer to quick and dirty techniques as ones which '. . . provide an approximate result or rapid feedback'. In practice this will usually mean taking an established evaluation method and then applying it in a less formal way than usual.

For example, a quick and dirty version of an experimental evaluation might be one in which some of the controls and balances were relaxed, in order to save time in the study design. Similarly, fewer participants would usually be involved than there would be in a more 'scientific' experiment and analysis might be limited to descriptive statistics only rather than any more complex statistical manipulations. A useful discussion of quick and dirty methodologies is given by Thomas (1996).

Investigator time

Sometimes the deadline will not be the limiting factor on the choice of evaluation method, but rather the time available to the investigator between the time at which an evaluation is commissioned and the time at which he or she must report back. For example, the deadline for producing the results of an evaluation may be a month away, but the investigator may only have ten days available in that time to work on the evaluation. In this sort of situation, one possibility is to choose a method where the burden on the investigator is low, even if the time to complete the evaluation may be longer.

The questionnaire is a method which comes into this category. Once a questionnaire has been designed, the investigator can send out as many as he or she wishes and then wait a couple of weeks for responses. This means, then, that the investigator would not have to run the evaluation in the same way as he or she would if conducting, say, an interview. For each hour of participant time in an interview, an hour of investigator time will also be taken up.

Another possibility would be to get someone else to carry out some part of the evaluation. For example, if an experiment or a think aloud protocol were to be used as the evaluation method, then it might be possible to have someone else conduct the trials. Similarly, it might be possible to hire market research groups to carry out interviews or run focus groups. This option, of course, presupposes the availability of human resources to support the investigator and the funds to hire these resources.

Participant time

Sometimes, especially when a product has a particularly narrow target user group, it can be difficult to find participants who have sufficient time to give to the study. Consider, for example, that the product in question were a medical system, with surgeons the intended user group. Clearly, it may not be reasonable to expect surgeons to take a lot of time out from a busy schedule in order to participate in a usability evaluation. In situations such as this the investigator should consider using a method where the time demands on the evaluation participant are comparatively low. This might mean, for example, that if a questionnaire or interview were to be used, it should be of the fixed-response variety rather than having open-ended questions. Fixed-response formats place less of a burden on participants because they simply have to make a selection from a number of categories or mark a scale in response to each question.

When the product under evaluation has already been released onto the market, one of the least costly evaluation methods in terms of participants' time is the field observation. This allows participants to carry on with their normal work whilst being monitored by the investigator in order to measure some aspect of their performance with the product. For example, if the product under test were a word processing package, the investigator might observe a secretary working with the package for a day to see if he or she has any problems with it. Provided the investigator is not too intrusive, this should not interfere too much with the secretary's work.

Money available

Often (indeed, in industry, usually – time and money amount to more or less the same thing) every hour that is spent on an evaluation has to be accounted for. Nevertheless, there are instances where the two are separable and where money can be constrained but time is not.

One way in which this might happen is with respect to the development of prototypes. Whilst there might be, for example, 15 days available to run an evaluation, there may not be enough money to build, say, an interactive software prototype to use in the evaluation. This, then, would rule out the use of methods which rely on the participant having interaction with the product, e.g. co-discovery, think aloud protocols and experiments.

Investigator knowledge

It may also be the case that the choice of method will depend on the level of expertise of the investigator with respect to the various methods. Some evaluation methods require extensive specialist knowledge and skills, whilst some require less. For example, in order to conduct an experiment the investigator must have an understanding of experimental design and the sorts of things that he or she must control and balance in order to ensure that systematic and misleading biases are eliminated from the data. Similarly, conducting an expert evaluation presupposes that the investigator really is an expert with respect to the principles of usable design.

Conversely, there are some methods that require comparatively little expertise to apply, or which require expertise only for parts of their application. The 'off-the-shelf' questionnaires discussed in Chapter 5 come with clear instructions as to how they should be administered and scored (although the interpretation of these scores in any particular context will require a level of expertise).

The private camera conversation (PCC) is an example of a method which requires expertise for only parts of its application. Expertise is required in designing the questions that participants will respond to and also to code and analyse what has been said. However, because the participant is talking to a camera rather than directly to the investigator the investigator does not need to have any expertise at interacting with participants in an evaluation session.

Of course, where the investigator does lack a specific skill, for example interviewing technique, it is possible to commission someone else to carry out that part of the evaluation. This may, however, be costly and presupposes that there is someone available to carry it out.

Participants available

In certain situations there may be few participants from the target group available to take part in the evaluation.

Consider, for example, that the target group for a product under test were men and women in high income professional careers. A high-end mobile phone might be an example of such a product. A problem that the investigator might face when recruiting participants from a user group such as this is that it may be difficult to persuade such people to take the time and effort to participate in the evaluation. Whilst, for example, the offer of, say, £20.00 to participate in an hour-long evaluation session might prove attractive to many, it is likely to be less attractive to someone who is well paid and under the time pressure that many in high income careers are.

In a situation where participants are difficult to recruit, the investigator may be best advised to use a non-empirical method or to use a method with which comparatively rich data can be gathered from only a small number of participants, e.g. think aloud protocols and co-discovery sessions.

Facilities and resources

The choice of method for a usability evaluation may also be affected by the facilities and resources available to the investigator. A facility that is becoming increasingly common in industry is the on-site usability laboratory (see Nielsen (1994) for examples). Usability laboratories are usually quiet rooms where evaluations can be conducted without the participants being disturbed by outside distractions, i.e. they provide a controllable environment. This is ideal for experiments and other methods where the participant interacts with a prototype or product, such as think aloud protocols or co-discovery methods.

Laboratories will often include audiovisual recording facilities, so that a record of the evaluation session can be kept for later analysis. Audiovisual resources are extremely helpful where an evaluation session is likely to generate more data than can reasonably be recorded by the investigator at the time. For example, with a focus group – assuming that the discussion is fairly lively – it would be impractical to expect that the investigator would be able to make a written record of all that was said in the focus group session itself. Therefore, at least an audio record should be made for later transcription and analysis. (In practice an audiovisual record might be preferable as this makes it easier to identify which member of the group said what.)

Audiovisual facilities can also be extremely useful in situations where participants are asked to perform unstructured interaction with a product – in other words, when the investigator has asked them to explore a product to see what it can do. This, again, is largely due to the difficulties involved in trying to make a written record of these actions as they occur. Methods that commonly involve asking the participant to explore an interface are, for example, think aloud protocols and co-discovery.

The human resources available to the investigator are also an important issue when considering the evaluation method to be used. This has already been mentioned above in the context of the possibility of asking others to conduct, say, interviews or experimental trials. Usually, where recordings are being made, the investigator will need the assistance of another person to operate the audiovisual equipment. Depending on the complexity of the situation and the quality of recording required, this person might need to have a fair degree of experience with audiovisual equipment. For example, if the product being evaluated were a software package running on a computer and the method being used were a think aloud protocol, there might be up to four different cameras recording at the same time. One might be focused on the participant's hands to monitor his or her key presses and mouse manipulations, another on the computer screen to monitor the effects of these inputs, one camera might be on the face of the participant to monitor his or her facial expressions and to make it clear when the participant was talking, whilst another camera could be on the investigator. As well as the visual recordings, it would be important to ensure that the voice of the investigator and participant came across on the recording.

If the recording were simply to be used for analysis purposes, then the requirements for picture and sound quality may not be particularly high. After all, the tape may not be seen by anyone other than the investigator, thus making a good impression on others would not be an issue. However, video is increasingly becoming a medium for presentation of evaluation results, in the form of summary videos that give an overview of 'typical' problems that participants had with a product and the attitudes that they expressed towards it. Here, then, a degree of professionalism in the picture and sound recording as well as in the quality of editing may give the outcomes of the evaluation more credibility in the eyes of those commissioning the evaluation and who watch the video. Thus, if this were the aim, the investigator would require the services of someone experienced with audiovisual recording to participate in the evaluation sessions.

Increasingly, contemporary ethics and practice in behavioural science are questioning the wisdom of conducting evaluations where the participant is left alone with the investigator as this can potentially put both the participant and investigator in a vulnerable position. This suggests that even in a situation where the investigator does not need any assistance he or she might be advised not to be alone with the participant. For many methods this is not an issue. For example, with questionnaires, incident diaries or logging techniques, the investigator need not be present at all. With field studies, there will normally be colleagues of the participants in the vicinity and with methods such as co-discovery, focus groups and user workshops there will be more than one participant present at a time anyway. However, if an evaluation method is being used which would usually involve the investigator and participant being alone in a one-to-one situation (e.g. experiment, interview, think aloud protocol), then the investigator might be wise to ensure that a third person was present to act as a 'chaperone'.

REPORTING THE EVALUATION

Clearly, it will be necessary to report the outcomes of the study to those who need to know. This will typically include, for example, product managers, designers, engineers, marketing personnel, etc. The way in which the evaluation study is reported is very important. It is important that reporting is clear and persuasive – otherwise it is possible that people will not understand what is being reported or that they will not be persuaded to act upon it.

Whilst all those involved in the product creation process will presumably have a common interest, in that they will all want to produce a product that is as good as possible, it is also possible that they may have partly conflicting interests. Perhaps the most common conflict of interests occurs between designers and engineers. Technical constraints set by engineers can be seen by designers as limiting the 'space' in which they have to work. At the same time, engineers may see some of the more 'imaginative' proposals put forward by designers as causing unnecessary technical complications.

It is important in reporting to have some understanding of the social dynamics going on between the interested parties and to report in a way which will not cause offence or attract accusations of unfairness from any party. A poor reporting style can lead to hurt feelings and hostility, with the result that the usability recommendations are ultimately less likely to be implemented than they would otherwise be.

Usually, evaluations are reported in one or more of three ways – written reports, verbal presentations and/or summary videos.

Written reports

These can be produced in varying degrees of formality depending on the purpose, the audience and the culture within the organisation. For full-length formal reports it is common practice to include the following sections:

Summary

This contains a summary of the purpose of the evaluation, the methodology followed, and the main results, conclusions and recommendations. This section should be about 100–150 words long, and is almost certainly the most important part of the report. It is the first thing that will be read and is thus the part of the report that will make the all-important first impression on the readers. Sometimes it may be the only part of the report that gets read at all – indeed a poorly written summary may be responsible for putting some people off reading the rest of the report, although others may only read the abstract purely due to time pressures on them. It is important, then, that the investigator includes in the summary the main message that he or she wants to communicate to the readers.

Introduction

The introduction will typically begin by introducing the objectives of the evaluation and the questions used. An objective might be, for example, to evaluate the usability of product X. Evaluation questions might be of the nature of '. . . What percentage of users could perform task Y without making an error?'. The introduction is also the part of the report where a description of the product under evaluation would be given, together with the background to the development of the product so far and how the evaluation contributed to this development. For example, details could be given about who needed to know the outcomes of the evaluation, what scope there would be for influencing the design of the product on the basis of the evaluation, and why an evaluation was being conducted at this particular stage of the product creation process.

Method

This section would include a description of the 'apparatus' used in the evaluation. This refers to what was evaluated and any equipment used in the evaluation itself. What was evaluated might be, for example, a finished product or some sort of product prototype or concept description. Equipment used might include, for example, audiovisual recording facilities, automatic logging devices or laboratory facilities.

The design of the evaluation would also be given in this section of the report, e.g. the evaluation methods that were used. This would be down to the level of, for example, listing the questions contained on a questionnaire and the tasks set in an experiment (although these might be relegated to an appendix which is referred to in the text). A description of the participants would also be included here, giving an overview of those characteristics which appear to be relevant in the context of the evaluation – this might include, for example, participants' ages and genders and their level of experience with the type of product being evaluated.

The method section also gives details of the procedure followed in the evaluation – this describes the evaluation sessions step by step from beginning to end. In the case of an experimental evaluation, for example, this would describe everything that happened from the moment that the participants turned up at the site to when they left, e.g. what the experimenter said to them before and after completion of the experimental tasks, the length of the session, how many breaks they had, how much they were paid for participation, etc. In the case of a survey method such as a questionnaire, the procedure would include describing how participants were chosen, contacted and asked to participate, how long it took to receive their responses, the return rate, how much respondents were paid, etc.

Results

This section gives the results of the evaluation in detail. This might include, for example, mean times on task or error rates, scores from attitude scales, or qualitative descriptions of problems observed or users' comments.

Discussion

This section is a discussion of the implications of the results, and also a discussion of the evaluation itself. Indeed, it is often advisable to begin this section with a discussion of the limitations of the evaluation – for example, to what extent is the sample that has been used really representative of the intended end-users, to what extent could the results be generalised, to what extent do problems found reflect what would happen when the product was used in the real world, etc?

The main part of the discussion, however, would be concerned with discussing the evaluation questions – given the results that have been found, what are the answers to these questions, what does this say about the adequacy of various aspects of the design and what are the implications.

Conclusions

This is a summary of the discussion. Concise answers to the evaluation questions are given, sometimes in the form of bullet points.

Recommendations

How can the outcomes of the evaluation be used to ensure that the product under creation (or the next version of the product) is optimal in terms of meeting the users' needs? Again, this may mean summarising issues that arose in the discussion section – bullet points are, again, sometimes used here.

Of course, the form of report delivered will depend on the nature of the evaluation carried out, the expectations of those who will be reading it and the extent to which the investigator is likely to have to defend the results. The full-length formal report described above would usually be written in circumstances where the evaluation conducted was fairly 'scientific', where those reading it expected the comprehensive reporting of the evaluation process and where the investigator might be expected to defend his or her work with respect to issues of methodology and analysis. Fully comprehensive reports are probably most commonly produced in academia and research establishments.

'Full' reports are less likely to be found in industrial or commercial contexts. A more common format in industry is the 'short report'. This can be anything from a slightly reduced version of the format described above to a one-page memo. Reduced versions of the formal report will probably begin with a summary, followed by a section on conclusions and recommendations. They may then have some detailed results and perhaps a brief outline of the method in appendices. The aim of a report such as this is to communicate the conclusions and recommendations to emerge from the evaluation whilst at the same time giving those who are interested an idea of the methodology followed. Usually, those reading the report will have little knowledge of behavioural science techniques or how evaluation studies of this nature should be designed, thus there is little point in going into the sort of detail that would be included in a full report.

A rule of thumb – in this author's experience – is that the better integrated usability issues are in an organisation, the shorter the reports produced by the usability specialist. Often, where usability issues have been very well established, a couple of pages will suffice, simply outlining the main results and giving conclusions and recommendations. This reflects an attitude from commissioners of '. . . we trust you to conduct an appropriate evaluation, now just tell us what we need to do to the improve the product'.

Verbal presentations

Again, these can vary in levels of formality. A formal presentation may involve talking in front of a room full of interested parties using visual aids such as slides and overhead sheets. This format would usually be a vehicle for a detailed explanation of all aspects of the evaluation, with a chance for the audience to ask questions. At the other end of the scale is the informal verbal presentation where the investigator might chat with others about his or her findings over coffee and biscuits. Here the structure of the presentation would be very loose. For example, the investigator might mention some of the main findings and recommendations and then see which direction the conversation went, depending on the questions that were asked. This informal approach usually presupposes that the audience have read an accompanying report or seen an accompanying video, whereas the more formal verbal presentation might be able to stand on its own (although, even with formal verbal presentations, it would be unusual if there were no accompanying written report, however short).

Probably the main advantage of verbal presentations is the opportunity they provide for the audience to interact with the investigator. This provides the opportunity for clarification of anything that is not understood, as well as for discussion of the conclusions and recommendations made. This can often be important in getting recommendations implemented. After all, with a written report, the commissioner might simply ignore any recommendations that he or she didn't find convenient and put the report in the filing cabinet, never to be seen again. However, it is not so easy to ignore recommendations made at a verbal presentation where the commissioner has the opportunity to challenge the investigator, giving the chance for issues to be properly 'thrashed out'.

Video presentations

Video presentations provide an excellent opportunity for the investigator to show an audience 'first hand' the sorts of problems that users experienced with an interface. This can be an extremely effective way to convince others of the significance of what was observed in an evaluation. For example, in a written or verbal presentation, the audience is, in a sense, expected to take what the investigator is reporting 'on trust'. When the investigator reports that users experienced difficulty with a particular task what are the audience to make of this? If the investigator describes something as a 'serious usability problem', then what does this mean for the users? The video presentation can answer questions such as these for the audience. There is often no substitute for showing users struggling with a product in order to convince others that changes to a design are needed.

Again, when video reports are used, it is usually in combination with a written (or perhaps verbal) presentation.

Getting evaluation findings implemented

One of the most common difficulties faced by those practising human factors is in getting their findings and recommendations implemented. So, how does the human factors specialist go about persuading those managing the product creation process to implement the findings of an evaluation?

As discussed above, convincing reporting of the evaluation is important. In particular the human factors specialist needs to communicate something of the human cost of the problems faced by users. Whilst being overdramatic should be avoided, it is important to be clear about the seriousness of the possible consequences of a usability problem. For example, in the case of products that are to be used in safety critical situations such as an in-car stereo, it may be that usability problems could lead to serious injury or even death of those using the product and – in this case – other road users. In other cases, of course, the consequences of usability problems will not be nearly so serious, yet could still be important. For example, with a video cassette recorder (VCR), usability faults are unlikely to put the user in any physical danger. Nevertheless, the faults might make it difficult to use the product for its intended purpose. Demonstrating problems such as these to those commissioning the evaluation will hopefully be enough to convince them of the need to make alterations. However, if they do not appear to be concerned about users' well-being, then some sort of evidence that usability problems are having an adverse effect on sales should do the trick!

Unfortunately – especially in the case of consumer products – this is easier said than done. Here, data linking specific usability problems with sales is notoriously difficult to gather in a valid and reliable way. The problem is that when asked, people often give 'rational' explanations for purchase decisions – these are, of course, likely to show a bias for buying products that have fewer usability faults. However in real life purchase decisions tend to be far less influenced by rational considerations. Nevertheless, as human factors specialists collaborate more and more with market researchers (e.g. Smit, 1996: Van Vianen, Thomas and van Nieuwkasteele, 1996), it may be possible to establish links between sales and particular usability issues. Certainly, industry now seems to accept a general link between usability issues and product sales (Jordan, Thomas and McClelland, 1996) – this is a driving factor behind the growing numbers of human factors specialists employed by firms manufacturing commercial products.

In the case of products that are designed for use in a professional setting these links are easier to make. If operators are producing more components per hour with a machine this is clearly likely to produce direct commercial benefits. Similarly if the operators were to make fewer errors and produce higher quality work with less wastage this will also bring direct financial benefits.

CASE STUDY: EVALUATING A SOFTWARE-BASED STATISTICS PACKAGE

So far in this chapter a number of factors that need to be taken into account when conducting an evaluation have been described and suggestions on reporting evaluations have been given. Chapter 5, meanwhile, outlined a number of methods that could be used for conducting evaluations. This case study illustrates how an evaluation could be conducted – taking into account the requirements, constraints, etc. of the evaluation and selecting appropriate evaluation methods. The (hypothetical) product under evaluation is a software-based statistics package.

Scenario

Assume that the manufacturers of this software package had one version of the package already on the market and were just under way with the development of a new package – so the product creation process was still at an early stage.

Purpose of the evaluation

It would be expected that in such a situation the manufacturers would want to have some general 'feel' for how usable their current package was – in other words to have a general overview of the current package's usability. They would also, presumably, want to know what the usability faults in the current package were, i.e. diagnostic information, and to know how these could be solved in the new design, i.e. prescriptive information. They would also want to know about these issues both in terms of performance with the product and attitudes towards the product.

Constraints and opportunities

Assume that the investigator had two working months to complete the evaluation in which he or she could spend one month of his or her own time. Imagine that there is an on-site usability laboratory available. Assume that the investigator were a human factors specialist with an academic background in psychology and that he or she was thus skilled with respect to the application of a wide range of techniques for the measurement of both performance and attitude. Assume also that he or she had access to a list of customers who held licences for this package.

Questions for the evaluation

The first step for the investigator would be to formulate questions for the evaluation which, when answered, would fulfil the purpose of the evaluation. In this case these questions might be along the lines of:

1   Can users perform a range of tasks with the current software within acceptable time limits and without making an unacceptably high number of errors?

2   Are there any usability faults in the product? If so, what are these?

3   How can these faults be put right?

4   What are users' attitudes towards this product?

5   What design changes are required to make users' attitudes more positive?

These, then, would be the five main questions that the evaluation would be addressing.

Approach to answering questions

The investigator would then have to consider the general approach that would be needed in order to answer each of the questions listed above. Examples of approaches that he or she might take are given below.

Question 1 – task performance

Firstly, the investigator would have to define a particular level of performance which could be regarded as being satisfactory. Perhaps he or she might say that acceptability should be judged by what users would expect and by comparison with other software packages on the market. He or she would then compare the level of performance of the current product against these expectations and against user performance with competing products.

Question 2 – usability faults

The investigator might tackle this question on the basis of observations of users interacting with the software and user reports of problems.

Question 3 – putting faults right

Here, the investigator might decide to suggest solutions on the basis of his or her observations, but also to take into account any comments from users.

Question 4 – users' attitudes

To answer this question the investigator might decide to quantify user opinion from a number of different dimensions and, perhaps, to gather some interesting quotes from users about their experiences of using the interface.

Question 5 – design changes

Again, the investigator might rely on a combination of his or her own judgement and a consideration of any comments from users.

Data needed

Having decided how the questions are to be answered, the investigator can then decide what data he or she needs to gather. This might be as follows:

Question 1 – task performance

Quantitative data about users' expectations with respect to the level of performance of the software. Quantitative data about level of performance with the current software and about performance with competitors' software.

Question 2 – usability faults

Quantitative data about user performance with various aspects of the product. Qualitative data based on investigator observation and user reports.

Question 3 – putting faults right

Qualitative data based on user reports and investigator observation.

Question 4 – users' attitudes

Quantitative user attitude data. Qualitative user attitude data.

Question 5 – design changes

Qualitative data based on investigator observation and user reports.

Methods to be used

Having decided on the data to be gathered, the investigator must then make a decision about which methods to employ in order to gather this data. This, of course, will not only depend on the data required, but also on the constraints and opportunities surrounding the evaluation.

Special opportunities

Perhaps the most appropriate starting point for the investigator would be to consider any special opportunities that he or she has in this situation which will help in gathering the required data. In this case, there appear to be two major advantages – the availability of the laboratory, which provides a controlled environment for gathering data, and the contact with the customers which provides the opportunity for access to people who use the current product.

Major constraints

Compared with many industry-based evaluations, the time constraints in terms of the deadline for the evaluation report and the investigator's time are not severe for an evaluation of this nature. The principal constraint in this situation is likely to be the time that the end-users of the product would be able to give to the evaluation. After all, this is a professional product, and so, almost by definition, the end-users are going to be working during the day. It seems unlikely that they would be able to give up large amounts of time to be involved with an evaluation such as this.

Methods

In any given situation, there may be many potential methods or combinations of methods that can be used for a usability evaluation. To some extent the combination chosen will always be dependent on the personal preferences of the investigator. However, given below is a four method combination which, given the requirements of this particular evaluation and the opportunities and constraints, should provide a comprehensive approach.

■ **Feature checklist:** The first step might be to employ a feature checklist in order to find out about the tasks that users most commonly perform with the software. The checklist, then, would contain a listing of the package's functionality. For each function, users could be asked to mark one of three boxes to indicate whether the function is something they use 'regularly', 'occasionally' or 'never'. This will give a feeling for the way the software is typically used, indicating which aspects of the package to concentrate on in subsequent parts of the evaluation.

   An advantage of feature checklists is that they do not take very long to complete – an important consideration in this context given that participants' time may be at a premium.

■ **Questionnaire:** After analysis of the feature checklist, the end-users of the current product could also be sent a questionnaire. This could contain questions pertaining to the attitudes towards the current software, expectations as to the level of performance that it should be possible to obtain with software of this type and questions giving users the opportunity to make any comments they wish to about the best and worst aspects of using the software.

   The investigator might well decide that the questions about attitudes towards the current product should be designed such that the attitudes could be quantified. Indeed, it might be appropriate to use an off-the-shelf attitude measurement questionnaire here, such as the Software Usability Measurement Inventory (SUMI) (Kirakowski, 1996) or the System Usability Scale (SUS) (Brooke, 1996).

   Having analysed the feature checklist, the investigator should now have an idea about the sorts of tasks that users do and/or would like to do with this type of software. On the basis of this, the investigator could include a list of tasks in this questionnaire and ask respondents how long they thought an acceptable time on task was for each of these, as well as asking them to indicate what sort of error rate they would find acceptable. Finally, there could be a number of open-ended questions asking about the best and worst aspects of the package, anything else that the respondents particularly liked or disliked about it and recommendations for improvement.

   As with the feature checklist, it is preferable to keep the questionnaire as short as possible. After all, the investigator is asking busy people to help out by responding, so it would be inappropriate to send them a questionnaire that took more than, say, 15 minutes to complete. If the SUS (which consists of ten questions with response scales) were used to measure attitudes, this should take around 3 or 4 minutes to complete. If respondents were asked about acceptable times and error rates for, say, 10 functions this might take another 5 minutes. If the questionnaire were to be completed within 15 minutes, this would leave another 6 or 7 minutes to respond to the questions about the best and worst aspects – hopefully this would be enough.

If it appears that the time to complete both the feature checklist and the questionnaire is too much to ask of the end-users, then the investigator could issue the feature
checklist to some people and the questionnaire to others. Clearly, this would require
double the number of participants if the response rate per instrument were to be
maintained. The time and effort involved for the investigator in finding these participants must, then, be weighed against how much time each of the participants could
reasonably be expected to give.

■ **Experiment:** One part of the evaluation could involve a laboratory-based experiment
to compare user performance with the current statistics package against user performance with a competitor's package. The other package used might be the market leader
or a package with a particularly good reputation for usability. The investigator would
set a range of tasks representative of the sorts of things that the package was typically
used for – he or she would know this from responses to the feature checklists. Time
on task and error rates could then be recorded and compared for one package against
the other. Similarly, these figures could act as a benchmark against which the new
product could be judged. For example, it might be decided that when the new product
was complete, users should be able to perform tasks making, say, 20% less errors and
spending an average of, for example, 20% less time on the tasks than with the best
performing package in this experiment.

Participants in this study would not be end-users of the current program – having
used the package would bias their performance with it. Also, the experimental session would be likely to last at least an hour, probably making it too long for most
of the 'real' end-users to participate in (especially when travel time to and from
the laboratory is considered). Probably the most practical solution would be to
invite students from the local university to participate and to pay them for their
time or to recruit people from a temporary employment agency. Because the packages being evaluated are statistics packages it would probably be appropriate only
to include participants with some understanding of statistics, for example social
science students.

If, as in many laboratories, this laboratory contains audiovisual recording equipment, then a record of the sessions can be kept. Some of this material might be used
in a subsequent video summary of the evaluation.

■ **Field observation:** Finally, the investigator could spend a day with a group of end-
users, observing how the package is actually used in practice and perhaps making an
audiovisual record of some of the users interacting with the package.

Conducting a field observation, even if very informally, will at least give the
investigator some feel for whether or not the issues that appeared to be important
under laboratory conditions actually appeared to be having much of an effect in the
real context of use. Similarly, the 'real life' incidents that have been captured on
video can later be played to others in the product development team to illustrate some
of the difficulties that users encounter.

Answering the research questions

Having conducted an evaluation using these four methods, how could the outcomes be
used to answer the research questions? Consider the questions one by one:

1. Can users perform a range of tasks with the current software within acceptable time limits and without making an unacceptably high number of errors?

This question could be answered by looking at the outcomes from the experiment and from the questionnaire. The experiment enables comparison of the current software package with that of the main competitor. If the performance of the current package were better or perhaps equal, in terms of time on task and error rate, this might be regarded as acceptable. If it were worse, this would almost certainly not be acceptable.

The performance of the current software could also be compared with the figures given on the questionnaire with respect to acceptable times and error rates for each of the set tasks. Admittedly, it is unlikely that the level of performance in the laboratory would be repeatable outside, given the distractions that occur in a real environment of use. Nevertheless, the performance data from the experiment should provide a useful indicator.

2. Are there any usability faults in the product? If so, what are these?

Again, this question could be answered primarily from the questionnaire and the experiment. The questionnaire provides the opportunity for users to note anything about the product that they do not like, whilst the investigator will also be able to spot problems when the participants have difficulties with any of the experimental tasks.

The investigator should also pick up on usability problems in the field study, even if this were conducted informally.

3. How can these faults be solved?

Here the investigator will have to draw mainly upon his or her knowledge and expertise with respect to designing for usability in order to find solutions to the faults observed in the experiment and field observation and the difficulties reported by users on the questionnaire. Whilst users are not designers, and thus cannot be expected to come up with practical design solutions to the problems, it would still be worth taking their suggestions for improvements into account when making recommendations to solve these faults.

4. What are users' attitudes towards the use of this product?

The questionnaire would be the main source of information on this point. By analysing responses to the attitude questions it is possible to draw up an 'attitude profile' for the product, showing the positive and negative attitudes that use of the product engenders.

5. What design changes are required to make users' attitudes more positive?

Looking at the answers to the open-ended part of the questionnaire would be a sensible starting point here. The aspects of a design that users cite as being amongst the worst aspects of the product and aspects of the product that they say cause them irritation are the ones that will need to be changed in order that the product elicits more positive attitudes from users.

The field observation is also likely to be important here. Comments that users make during these sessions will probably show what aspects of the design they are particularly displeased with. These, and the aspects cited on the questionnaire, can then be altered within the framework of the usable design principles outlined in Chapter 3.

Reporting the study

The main deliverable from a study such as this would probably be a report outlining the evaluation questions, the procedure followed, and the results. The report would usually include a discussion of these results and then summarise the conclusions that the investigator had drawn. Finally, recommendations would be made. These would usually include recommendations for design features that should be incorporated into the new design. However, they might also include recommendations about, for example, the product creation process that should be followed in developing subsequent versions of the product or even recommendations about how future usability evaluations should be conducted based on the investigator's experience of this evaluation.

It would also be a good idea in this case to create a video summary of the evaluation using material shot in the field observation and the experiment. What can be particularly effective in such situations is to show a clip of some task that caused particular difficulty in the experiment and then show a clip of somebody struggling with it in the field observation. As well as giving a graphic illustration of a problem that users encountered, this is also a way of defending the experiment against the criticism that effects found in the laboratory are of no real significance in the 'real world'.

CHAPTER SEVEN

Conclusions

In this introductory text, an overview of usability and related issues has been given. In the first three chapters, the concept of usability was introduced and defined and the basic principles of designing products for usability were given. Chapter 4 described how to incorporate usability issues into the design process, whilst Chapters 5 and 6 concentrated on usability evaluation.

Usability is a 'hot' issue at the moment. As mentioned at the beginning of the book, users are demanding usable products and manufacturers are beginning to see designing for usability as bringing a competitive advantage. Certainly, usability is a commercial issue. However, it is more than just that. Providing adequate standards of usability is part of the responsibility that those producing products owe to those who use them. We live in a high-technology era. We live with technology that has the power to enrich our lives – in our homes, in our workplaces and in our communities. Yet all too often this technology proves difficult for us to control. The consequences of this can range from mild annoyance to serious injury or death. Usability is about giving control back to the user – producing useful, pleasurable, safe and helpful products that will enhance the quality of our lives.

References

ALLEN, R.B. and SCERBO, M.W., 1983, Details of command language keystrokes, *ACM Transactions on Office Information Systems*, **1** (2), 159–78.

ALLWOOD, C.M., 1984, Analysis of the field survey, in Allwood, C.M. and Lieff, E. (Eds), *Better Terminal Use*, pp. 72–7, University of Goteberg: Syslab-G.

BEAGLEY, N.I., 1996, Field based prototyping, in Jordan, P.W. *et al.* (Eds), *Usability Evaluation in Industry*, pp. 95–104, London: Taylor & Francis.

BONNER, J.V.H. and CADOGAN, P., 1991, Important issues in successfully using 'quick and dirty' methods in ergonomics design consultancy, in Lovesey, E.J. (Ed.), *Contemporary Ergonomics 1991*, pp. 381–6, London: Taylor & Francis.

BROOKE, J., 1996, SUS – A quick and dirty usability scale, in Jordan, P.W. *et al.* (Eds), *Usability Evaluation in Industry*, pp. 189–94, (London: Taylor & Francis).

CARD, S.K., MORAN, T.P. and NEWELL, A., 1983, *The Psychology of Human Computer Interaction*, Hillsdale, New Jersey: LEA Publishers.

ECONOMIST, 1995, *The Pocket World in Figures*, London: The Economist.

EDGERTON, E.A., 1996, Feature checklists: a cost effective method for 'in the field' evaluation, in Jordan, P.W. *et al.* (Eds), *Usability Evaluation in Industry*, pp. 131–7, London: Taylor & Francis.

EDGERTON, E.A. and DRAPER, S.W., 1993, 'A comparison of the feature checklist and the open response questionnaire in HCI evaluation', *Computing Science Research Report* GIST-1993-1, University of Glasgow: Department of Computing Science.

GIBSON, W.H. and MEGAW, E.D., 1993, An ergonomic appraisal of the Piper Alpha disaster, in Lovesey, E.D. (Ed.), *Contemporary Ergonomics 1993*, pp. 202–7, London: Taylor & Francis.

GRUDIN J., 1989, The case against user interface consistency, *Communication of the ACM*, **32** (10), 1164–73.

HART, S.G. and STAVELAND, L.E., 1988, Development of the NASA-TLX (Task Load Index): Results of empirical and theoretical research, in Hancock, P.A. and Meshkati, N. (Eds), *Human Mental Workload*, pp. 139–83, North Holland: Elsevier.

HARTEVELT, M.A. and VIANEN, E.P.G., VAN, 1994, User interfaces for different cultures: a case study, in *Proceedings of Human Factors and Ergonomics Society Conference 1994*, pp. 370–3, California: Human Factors and Ergonomics Society.

HOWES, A. and PAYNE, S.J., 1990, Display-based competence: towards user models for menu-driven interfaces, *International Journal of Man–Machine Studies*, **33**, 637–55.

JORDAN, P.W., 1992a, Claims for direct manipulation interfaces investigated, *Industrial Management and Data Systems*, August, 3–6.

JORDAN, P.W., 1992b, Ergonomic design for vehicle users of new technology interfaces, *Engineering Designer*, May, 16–20.

JORDAN, P.W., 1992c, Creeping featurism is the challenge for safer in-vehicle interfaces, *Automotive Engineering*, **17** (1), 34–5.

JORDAN, P.W., 1993, Methods for user interface performance measurement, in Lovesey, E.J. (Ed.), *Contemporary Ergonomics 1993*, pp. 451–60, London: Taylor & Francis.

JORDAN, P.W., 1994a, What is usability?, in Robertson, S. (Ed.), *Contemporary Ergonomics 1994*, pp. 454–8, London: Taylor & Francis.

JORDAN, P.W., 1994b, Focus groups in usability evaluation and requirements capture: a case study, in Robertson, S. (Ed.), *Contemporary Ergonomics 1994*, pp. 449–53, London: Taylor & Francis.

JORDAN, P.W. and JOHNSON, G.I., 1991, The usability of remote control for in-car stereo operation, in Lovesey, E.J. (Ed.), *Contemporary Ergonomics 1991*, pp. 400–7, London: Taylor & Francis.

JORDAN, P.W. and JOHNSON, G.I., 1993, Exploring mental workload via TLX: the case of operating a car stereo whilst driving, in Gale, A. *et al.* (Eds), *Vision in Vehicles IV*, pp. 255–62, North Holland: Elsevier.

JORDAN, P.W and MOYES, J., 1994, Does icon design really matter? in Robertson, S. (Ed.), *Contemporary Ergonomics 1994*, pp. 459–64, London: Taylor & Francis.

JORDAN, P.W. and O'DONNELL, P.J., 1992, The Index of Interactive Difficulty, in Lovesey E.J. (Ed.), *Contemporary Ergonomics 1992*, pp. 397–402, London: Taylor & Francis.

JORDAN, P.W. and SERVAES, M., 1995, Pleasure in product use: beyond usability, in Robertson, S. (Ed.), *Contemporary Ergonomics 1995*, pp. 341–6, London: Taylor & Francis.

JORDAN, P.W. and THOMAS, B., 1995, . . . But how much extra would you pay for it? An informal technique for setting priorities in requirements capture, in Robertson, S. (Ed.), *Contemporary Ergonomics 1995*, pp. 145–8, London: Taylor & Francis.

JORDAN, P.W., McCLELLAND, I.L. and THOMAS, B., 1996, Introduction, in Jordan, P.W. *et al.* (Eds), *Usability Evaluation in Industry*, pp. 1–3, London: Taylor & Francis.

JORDAN, P.W., THOMAS, B. and McCLELLAND, I.L., 1996, Issues for usability evaluation in industry: seminar discussions, in Jordan, P.W. *et al.* (Eds), *Usability Evaluation in Industry*, pp. 237–43, London: Taylor & Francis.

JORDAN, P.W., DRAPER, S.W., MACFARLANE, K.K. and McNULTY, S.-A., 1991, Guessability, learnability and experienced user performance, in Diaper, D. and Hammond, N. (Eds), *People and Computers VI*, pp. 237–45, Cambridge: Cambridge University Press.

KEMP, J.A.M. and GELDEREN, T. VAN, 1996, Co-discovery exploring: an informal method for iteratively designing consumer products, in Jordan, P.W. *et al.* (Eds), *Usability Evaluation in Industry*, pp. 139–46, London: Taylor & Francis.

KERR, K.C. and JORDAN, P.W., 1994, Evaluating functional grouping in a multi-functional telephone using think-aloud protocols, in Robertson, S. (Ed.), *Contemporary Ergonomics 1994*, pp. 437–42, London: Taylor & Francis.

KERR, K.C. and JORDAN, P.W., 1995, An investigation of the validity and usefulness of a 'quick and dirty' usability evaluation, in Robertson, S. (Ed.), *Contemporary Ergonomics 1995*, pp. 128–33, London: Taylor & Francis.

KIRAKOWSKI, J., 1996, The software usability measurement inventory: background and usage, in Jordan, P.W. *et al.* (Eds), *Usability Evaluation in Industry*, pp. 169–77, London: Taylor and Francis.

LANDAUER, T.K., 1987, Relations between cognitive psychology and computer system design, in Carroll, J.M. (Ed.), *Interfacing Thought*, pp. 1–25, London: MIT Press.

MAISSEL, J., 1990, Development of a Methodology for Icon Evaluation, *NPL Report*, DITC 159/90, Teddington, UK: National Physical Laboratory.

MAYES, J.T., DRAPER, S.W., McGREGOR, A.M. and OATLEY, K., 1988, Information flow in a user interface: the effect of experience and context on the recall of MacWrite screens, in Jones,

D.M. and Winder, R. (Eds), *People and Computers IV*, pp. 275–89, Cambridge: Cambridge University Press.

McCORMICK, E.J. and SANDERS, M.S., 1983, *Human Factors in Engineering and Design*, 5th Edn, Auckland: McGraw-Hill.

MOYES, J. and JORDAN, P.W., 1993, Icon design and its effect on guessability, learnability and experienced user performance, in Alty, J.L. *et al.* (Eds), *People and Computers VIII*, pp. 49–59, Cambridge: Cambridge University Press.

NIELSEN, J., 1994, Special Issue on Usability Laboratories, *Behaviour and Information Technology*, **13** (1 and 2).

NORMAN, D.A., 1988, *The Psychology of Everyday Things*, New York: Basic Books.

NORMAN, D.A., DRAPER, S.W. and BANNON, L.J., 1986, Glossary, in Norman, D.A. and Draper, S.W. (Eds), *User Centred System Design*, pp. 487–97, Hillsdale, New Jersey: LEA Publishers.

O'DONNELL, P.J., SCOBIE, G. and BAXTER, I., 1991, The use of focus groups as an evaluation technique in HCI, in Diaper, D. and Hammond, N. (Eds), *People and Computers VI*, pp. 211–24, Cambridge: Cambridge University Press.

PAYNE, S.J. and GREEN, T.R.G., 1986, Task-action grammars: a model of the mental representation of task languages, *Human-Computer Interaction*, **2**, 93–133.

PHEASANT, S., 1986, *Bodyspace: Anthropometry, Ergonomics and Design*, London: Taylor & Francis.

RAVDEN, S.J. and JOHNSON, G.I., 1989, *Evaluating Usability of Human–Computer Interfaces: a Practical Method*, Chichester: Ellis Horwood.

REISNER, P., 1990, What is inconsistency? in Diaper, D. *et al.* (Eds), *Human Computer Interaction – INTERACT '90*, pp. 175–81, North Holland: Elsevier.

RIJKEN, D. and MULDER, B., 1996, Information ecologies, experience and ergonomics, in Jordan, P.W. *et al.* (Eds), *Usability Evaluation in Industry*, pp. 49–58, London: Taylor & Francis.

SHACKEL, B., 1986, Ergonomics in design for usability, in Harrison, M.D. and Monk, A. (Eds), *People and Computers: Designing for Usability*, pp. 44–64, Cambridge: Cambridge University Press.

SHACKEL, B., 1991, Usability – context, framework, definition, design and evaluation, in Shackel, B. and Richardson, S. (Eds), *Human Factors for Infomatics Usability*, pp. 21–37, Cambridge: Cambridge University Press.

SMIT, E., 1996, Usable usability evaluation: if the mountain won't come to Mohammed, Mohammed must go to the mountain, in Jordan, P.W. *et al.* (Eds), *Usability Evaluation in Industry*, pp. 19–28, London: Taylor & Francis.

THOMAS, B., 1996, 'Quick and dirty' usability tests, in Jordan, P.W. *et al.* (Eds), *Usability Evaluation in Industry*, pp. 107–14, London: Taylor & Francis.

VERMEEREN, A.P.O.S., 1996, Getting the most out of 'quick and dirty' usability evaluation, in Jordan, P.W. *et al.* (Eds), *Usability Evaluation in Industry*, pp. 121–8, London: Taylor & Francis.

VIANEN, E. VAN, THOMAS, B. and NIEUWKASTEELE, M. VAN, 1996, A combined effort in the standardisation of user interface testing, in Jordan, P.W. *et al.* (Eds), *Usability Evaluation in Industry*, pp. 7–17, London: Taylor & Francis.

VRIES, G. DE, HARTEVELT, M. and OOSTERHOLT, R., 1996, Private Camera Conversation method, in Jordan, P.W. *et al.* (Eds), *Usability Evaluation in Industry*, pp. 147–55, London: Taylor & Francis.

Index